DNA from A to Z & Back Again

DNA from A to Z & Back Again

Carol A. Holland, PhD, MT (ASCP)
Daniel H. Farkas, PhD, HCLD

A Primer on Genetics and Molecular Medicine for . . . . Everyone

◆

WITH A POINT/COUNTERPOINT ON DIRECT-TO-CONSUMER GENETIC TESTING: IS IT FOR YOU?

The Fully Updated Version of the Popular *DNA from A to Z*

1850 K Street, NW, Suite 625
Washington, DC 20006

Cover design by Rick Nance.

1 2 3 4 5 6 7 8 9 0 VRS 10 09 08

Printed in the United States of America

Library of Congress Cataloging-in-Publication Data

Holland, Carol A.

DNA from A to Z & back again: a primer on genomics & molecular medicine for . . . everyone / Carol A. Holland, Daniel H. Farkas.

p. cm.

ISBN 978-1-59425-088-0

1. DNA—Dictionaries. 2. Molecular diagnosis—Dictionaries. 3. Genomics—Dictionaries. I. Farkas, Daniel H. II. Title.

QP624.H635 2008
572.8'603—dc22

2008022356

To Steve, my brother, who succumbed to cancer in 2007 on his 61st birthday. His disease was related to many of the concepts in this book, including a last-ditch effort near the end to utilize genomic profiling and personalized medicine for proper therapy. I hope his story (I will not supply the details; as we age we will all unfortunately become all too familiar with a similar story, equally intimately) spurs you to learn more about these concepts so that you can become a more active participant in your healthcare and that of your loved ones. Steve encouraged me to write about what I know; for that I'll always be thankful. He told me repeatedly he was proud of me; his courage and accomplishments were why I was even more proud of him.

D.H.F.

To my father, Jack, and to the memory of my mother, Minervia. She would have loved this book.

C.A.H.

Contents

Preface

Progress in DNA science, biology, and genomics is so rapid that it's time for another edition of the popular *DNA Simplified: The Hitchhiker's Guide to DNA*. First published in 1996, the illustrated version and second edition came out a short two years later, reflecting the rapid growth rate in genomic medicine. The book was revised in 2004 and re-named *DNA from A to Z.*

Genomic advances and their influence on medicine continue to outpace regulators' ability to deal with the growing technology and information base and with the marketplace's ability to assimilate all the opportunities seamlessly. In time, this too shall pass and genomic medicine will find its niche in American and global capitalism and medicine. A preface is no place to editorialize, so we won't, except to say that it's been a great ride in molecular diagnostics and what is "morphing" into personalized and genomic medicine.

A word about the title, specifically the phrase "… & Back Again": When molecular diagnostics was young, a scant two decades ago, it was ahead of the curve, introducing new technologies and clinical laboratory applications to various diseases. Physicians, regulators, and insurers needed time to assimilate these technologies and applications into the practice and business of medicine. Many so-called "molecular" (or "DNA-based") tests have now become the standard of care-and heavily relied upon by physicians as well as by criminal investigators and attorneys (think HIV viral load or DNA-based forensic and paternity testing). But the sequencing of the human genome was completed only five short years ago, in 2003. New tools (see "DNA chips") abound. It's only natural that new applications (profiling tumors' DNA, stem cell research, cloning, etc.) arise out of these advances. So the more established portion of the field is mainstream, and yet those of us at the cutting edge of 21st-century genomics are discovering and implementing even more advanced applications that physicians, regulators, and insurers again need time to assimilate into the practice and business of medicine today. For those of us in the field, it's déjà vu all over again, as a famous baseball philosopher once suggested.

So with all of the above being true, and because so many professionals and lay people alike are interested in DNA and genomics (and why not? it's all about the most personal aspects of our being, our DNA and its effect on our health and the health of our descendants), I (DHF) decided it was time for a fourth edition of the original. The field has advanced so much so quickly that it has led me personally to an exciting new venture in genomics; I was too busy to undertake this alone and so I am excited and honored that I was able to recruit a coauthor to help with this edition. Dr. Carol Holland is a friend and colleague but more importantly for you, the reader, her expertise and reputation in the field make her a perfect coauthor. Her contributions to this book are excellent and provide new insights and a fresh look for the text. Dr. Holland's *bona fides* are well documented in her "bio" at the end of the book.

As the book's subtitle (*A Primer on Genomics for Everyone*) suggests, this book has broad appeal. It is a primer, a glossary, a jumping-off point for the reader to learn more about any subject in this book. Yet a printed book cannot possibly keep pace with the amount of information emerging on the Web. You could "Google" any term in our greater than 100 pages of entries and (in 0.03 seconds) hit upon hundreds, thousands, tens of thousands of entries. (Esoteric humor: early in the history of the field, molecular diagnosticians used to joke that the power of PCR was its sensitivity while the weakness of PCR was ... its sensitivity. Today, the power and weakness of the Web is its breadth; how do you know which of those hundreds or thousands of "hits" is worth reading?)

My hope, and Carol's too, is that this book will serve as a guide or starting point for your search and help you sift through the vast amount of data on the Web. After all, it's your DNA, your health, your most personal questions you're seeking to investigate. Why not have a solid foundational understanding of the terms?

You'll appreciate this book if you are

- A medical school student ... or a high school student.
- A Wall Street biotech analyst ... or a diagnostics company sales rep.
- A patient trying to figure out where to start ... or a healthy individual who wants to stay that way (see Point/Counterpoint on Direct-to-Consumer Genetic Testing that follows).
- A hospital administrator interested in what's going on in that lab across the parking lot ... or a hospital administrator trying to understand where new revenue opportunities may originate.
- A Congressional aide seeking a handle on the hottest new area of healthcare ... or an insurance executive seeking clarification on new and seemingly expensive healthcare solutions.

In other words, everyone's interested in DNA on one level or another because we're all interested in our health. Francis Collins, director of the National Human Genome Research Institute, has suggested that every leading cause of death with the possible exception of accidents (there's a subtle bit of dark humor in the use of the word "possible") has a genetic component.

A note about the text—in an effort to make the book as broadly appealing as possible, we have targeted the interested layperson, who is not familiar with most or any of the terms in this book, as our audience. And we've included cross-references (noted with the symbol "☞") to provide links to additional information. More knowledgeable or technically savvy professionals may also have reason to pick up this book and for them we have included entries marked "TECHIE:" for "TECHNICAL NOTE" that they may find interesting or useful. That means the rest of you can skip over those sections, but if you read and don't understand them-that's OK!

So enjoy the book. Carol and I guarantee that you or someone you know will find it useful.

Daniel H. Farkas, PhD
Carol A. Holland, PhD
Somewhere near Detroit
May 2008

Direct-to-Consumer Genetic Testing: Is It for You?

PRO DTC Testing, by Daniel H. Farkas:

I'm going to "out" myself. I have bouts of pruritis (itchy skin). During one maddening attack and before the antihistamine I took exacted its drowsy effect, I searched the Web for some answers. WebMD.com touts itself as "[t]he leading source for trustworthy and timely health and medical news and information [on the Web]." That last little bracketed part is important and I have added it with literary license because I assume the authors meant to communicate that little tidbit.

Amazingly, even to someone in the field, the site informed me about a gene, *GRPR*, which when mutated or knocked out in mice made these mice more resistant to itching after topical chemical insult. Normal mice scratched much more vigorously upon insult than *GRPR*-deficient mice, who still scratched but less so, suggesting a multigenic response to itching (1).

Interesting? Surely. A cure for my pruritis? Hardly. But I knew this because of my training; to me words like "mouse," "target for new drugs," and "researchers" instantly alerted me to the fact that anything of value to me personally from these finding was perhaps a decade away. I was motivated to find solutions elsewhere.

But what of literate yet unscientific readers? Their recognition of the ultimately identical conclusion would come more slowly than mine did, but it would be the same.

Is there harm here? Is there false hope? I would argue that neither exists. Indeed, our free press, which reports most new "gene findings" in sensationalistic terms, serves the vital function of educating readers about genetics, genomics, and the future of medicine. Pruritis is one thing; what if the press release concerns something more serious, perhaps cancer or Huntington's disease? Disappointment might be greater but the process remains the same: the notoriety provides those interested the opportunity to learn more and draw their own conclusions about the nature of research in Western civilization and how long it takes to lead to therapies (much longer than it takes to lead to new diagnostics, which also have much value).

Let's explore that process of lay readers' realization a bit more. They'll learn of the finding and poke around the Web. They'll come across, I hope, mostly scholarly pursuits of the subject. They'll learn in relatively short order, or perhaps after many grueling hours of reading, that a press release or gene finding is not tantamount to cure or therapy. And if it's something as benign as pruritis, they'll leave it at that and continuing searching for the right skin moisturizer.

What if it isn't something benign? What if the reader is the patient or the loved one of a seriously ill patient? The reader will most definitely not "leave it at that." He or she will exhaust every option for education about the matter and instead of hours and a private Web search, will ultimately involve professionals who will be given the opportunity to explain the facts of the situation to the inquirer.

Is any of this bad or harmful? I think no one would argue that. To the contrary, it gets people reading about and interested in genomics and its promise. Knowledge is power and helps patients cope, moves clinical trials forward, helps spur fundraising, and leads to all manner of other "motherhood and apple pie" outcomes.

So was it irresponsible or "bad" for Washington University to issue a press release in July 2007 about *GRPR*, even if a cure or therapy for pruritis remains years away? No. The university wanted to bring appropriate notoriety to the issue and its fine institution. The public became aware and this awareness advanced the field, if by a tiny increment. Substitute amyotrophic lateral sclerosis (ALS, also known as Lou Gehrig's disease) for pruritis in a hypothetical press release; would that have been problematic? No … ALS patients' hopes would have soared and then been dashed upon ultimately learning of how the process works, but they would have become more informed; they would have become consumers of genomics information. With information can come understanding, improved communication, and progress, even if that progress is often painfully and glacially slow.

So what does all of the above have to do with a position piece in favor of direct-to-consumer (DTC) genetic testing? I submit that it is no different, with one key exception (money, what else?), from lay people learning about a genomic finding on WebMD.com. If I didn't suffer from pruritis, I might never have learned about *GRPR*. Now I know; that's a good, if benign, thing. And if I was researching something more serious, for example, breast cancer, and was concerned about my daughter's likelihood of having inherited a mutation in *BRCA*-1 or -2, what then?

In 2008, and for some years now, I have had an option to obtain that information. The exception mentioned above now rears its ugly head, namely, the cost of the information. Unlike information at WebMD.com, knowledge of one's *BRCA*-1 and -2 sequences is not free. But is the fact that this specific knowledge costs thousands of dollars "ugly"? I don't think so; no one, with the exception of my family, has the right to counsel me on how to spend my money. If I, and by extension the marketplace, value knowledge of a gene's sequence at thousands of dollars, no one has the right to tell me I can't use this service, if I'm willing to pay for it. That includes regulators not having the right to stand in my way indirectly by putting regulatory hoops in front of the provider (though as a hypothetical consumer of this health information I am delighted that regulatory bodies exist to quash "quackery" and promote quality). If the clinical laboratory is not up to appropriate standards; if the information is not vetted by a credentialed clinical laboratory director; if it is not communicated with knowledge, professionalism, and sensitivity by qualified physicians and genetics counselors, the marketplace will quickly weed out the weak, and only the fittest of services will survive. Furthermore, safeguards against actionable information provided by DTC genomic companies exist. Physicians are appropriate gatekeepers; a recommended unusually early colonoscopy or prophylactic mastectomy, to name but two examples of possible outcomes of DTC genomic information, are well "regulated" by the inevitable involvement of the patient's physician or surgeon.

Direct-to-consumer genetic testing is here. With advances in technology, the completion of the sequencing of the human genome, and the pressures of capitalism, its arrival was inevitable. Are there some "snake oil salesmen" out there? Maybe. Are there also responsible, professional enterprises providing well intentioned, appropriately communicated information to consumers? Certainly. Is the information of value? Is the information worth the price of acquisition? That's not only a question for the marketplace but one we can rest assured the marketplace will settle long before we academics reach consensus or establish guidelines, and long before regulators define jurisdiction and issue pronouncements. In the meantime, it falls to the consumer to employ the wisdom found in the Latin phrase "*caveat emptor*," so that s/he is not a victim of an equally famous note on capitalism: "There's a sucker born every minute." (Some have suggested P. T. Barnum really meant that "There's a customer born every minute." That's certainly a much less malignant sentiment.)

The French philosopher Voltaire, who influenced many important thinkers of the French and American revolutions, made a point highly relevant to this debate. He famously wrote, "I disapprove of what you say, but I will defend to the death your right to say it." As molecular diagnosticians and physicians (2), we may (or may not) disagree with the value and process of disseminating genomic information provided directly to consumers, but we will fail if we stand in the way of the consumer's right to acquire it. Indeed, we advance our own standing and reputations if we assist in the marketplace by sharing our protocols, teaching quality control, and removing misunderstanding and irrational fear of genomic information.

References

1. Sun YG, Chen ZF. A gastrin-releasing peptide receptor mediates the itch sensation in the spinal cord. Nature 2007;448,700–3.
2. Hunter DJ, Khoury MJ, Drazen JM. Letting the genome out of the bottle—will we get our wish? N Engl J Med 2008;358;105–7.

CON DTC Testing, by Carol A. Holland:

Normally Dan and I pretty much agree on everything. This topic, though, is one on which we do not. The question should not be whether people should be proactive in decisions regarding their own healthcare. Of course they should be. The question should be whether "direct-to-consumer" (DTC) genetic tests are beneficial in making those decisions.

The "completion of sequencing the human genome" was highly publicized; however, many people misinterpreted that milestone to read the "completion of *knowing everything possible about* the human genome." In actuality, while we have come a very long way, we still have a long way to go to fully understand the relationship between genes and disease. Any scientist studying genetics will freely admit this. Recently, though, a number of companies have been exploiting the genetic association between disease and health by offering misleading DTC genetic testing.

As Dan pointed out, the internet has allowed easy access to a ton of information. For the most part, this is a good thing. However, as we know, not all internet sites are created equal. One study performed by the University of Michigan found that the information on internet sites set up specifically to address melanoma, a type of skin cancer, contained wrong (14%), missing (62%), or misinterpreted (41%) information (1).

The assumption that people will thoroughly research a topic and ultimately come to the right conclusion is questionable. Take antibiotics, for example; most of us know that antibiotics are only effective against bacterial infections. All too often, however, we find that patients with viral infections will demand an antibiotic from their physician, and the physician relents and writes a prescription for the sole purpose of keeping the patient happy. (This, by the way, has enormously added to the antibiotic resistance crisis.)

The other assumption is that people will research and analyze the scientific data accurately and correctly. Some may, but most probably will not. Take the much publicized case where Tom Cruise criticized Brooke Shields for taking antidepressants for postpartum depression. During his talk with Matt Lauer on the Today show, when he criticized Ms. Shields, the topic turned to the drug Ritalin (2). Mr. Cruise stated "... you have to evaluate and read the research papers on how they came up with these theories ... that's what I've done ... ." He may have read them, but he did not analyze them correctly. I would have had a hard time analyzing them because they are not in my field of interest and I am at least a scientist. The problem is that there is a vast amount of information available, and analyzing the information is tantamount to doing a doctoral-level research project.

Interesting, but how does all this relate to DTC genetic testing?

First, I need to define different types of DTC genetic testing. One type is directed toward an individual who has a family member afflicted with a specific disease. The individual is concerned that s/he may have inherited that disease. The individual does an internet search, then finds a mutation associated with that disease and a laboratory that will test for that mutation without a doctor's order. If the individual learns s/he does not have the mutation, all is seemingly well. But what if the test indicates that the individual did inherit the mutation—what then? If the individual shares the results with his/her physician, the physician will probably repeat the test using a properly credentialed laboratory whose director s/he knows, trusts, and has used before. If the "consumer" of genetic testing chooses not to share the results with his/her physician (reasons could include not wanting the insurance company or employer to find out), the physician is at a disadvantage in providing proper or new treatments.

So, in this case, what is the problem with DTC genetic testing? The person researched the topic and paid for the test, so is there any harm done? There may or may not be, depending on the scenario. What if the test came back "negative" and the patient started to develop symptoms? What if there were other genes or mutations associated with the disease? What if the test was not performed correctly because the lab was not properly certified? What if other preventable factors play a role in disease development? In any of these scenarios, patients would have access to the best possible treatment only if the test would have been performed under a physician's care. By choosing DTC testing, patients are placing themselves in roles they are neither equipped nor educated enough to handle.

There's another type of DTC genetic testing; it's equivalent to genetic fortune telling. Several companies have recently offered DNA analysis and made a wide variety of claims that range from the ability to determine ancestry to choosing cosmetics based on DNA sequence (we can do the first; we cannot do the second). These examples are fairly harmless, and if people choose to spend their money for these purposes, that's their prerogative. But what about companies that make claims regarding genetic sequences and health?

As I was researching this topic, I came across an article by The Associated Press: "Home Bipolar Disorder Test Causes Stirs" (3). One of the most troublesome parts of the article is this quote: "[Kelsoe] acknowledges that bipolar disorder probably results from a combination of genetic factors and life experiences, and that the presence of these gene variations does not at all mean that someone will, in fact, develop the disease. He admits, too, that his findings about the genetic basis of the illness are far from complete." Yet, after his own admission, he still promotes this test—all while collecting a large fee. This sounds like a conflict of interest and not necessarily one that is in the best interests of the consumer.

Scientists, and probably most of you interested in reading this book, know that while some diseases have a clearly associated mutation—i.e., if the mutation is present, the disease manifests itself (it should be stressed that a physician is still required for disease management)—most diseases do not have such a direct "cause and effect" relationship. Many, many additional factors such as eating habits, exercise, smoking, weight, family history, environmental factors, ethnic background, and other health conditions play major roles in whether or not an individual will get a disease. Still, companies manipulate genetic data and make claims they know are not true. Most consider this "scientific fraud."

Many companies that offer this type of testing do not interpret the results in the context of true clinical validity or have genetic counselors on hand to interpret and communicate the results appropriately. In fact, a quick internet search showed that many DTC genomics companies' board members have little or no scientific training, certainly not advanced genetics training. Although the advisory boards did consist of doctoral-level individuals, degrees ranged from genetics to mathematics to computer science. In addition, the geneticists were research focused and I did not discover any that were board certified in human genetics. Still, such companies promote genetic testing; but how can they do this?

These companies are not (currently) regulated by the FDA or Federal Trade Commission (FTC) because they do not call what they do "testing." This is the molecular genetics equivalent to the health product industry where we have all seen claims similar to "we guarantee you will lose up to 20 pounds in just 10 days."

OK, you've paid your money, gotten the test and received a result; what does it all mean? Because the companies are not allowed to make diagnoses, the accompanying "report" focuses on questionable statistics and recommendations about changing lifestyles, eating habits, etc.; in other words, advice that is generic or at the very least given freely (with an emphasis on free) by family members, neighbors, and friends.

For most individuals, a very detailed disclaimer on the forms that one must sign prior to testing should present a very strong clue that this is all fraught with problems. Would anyone buy a car that required signing a form saying that there was absolutely no responsibility to the company once the check was cashed? Would one hire contractors to build an addition to one's home if there wasn't any oversight or warranty? With health being so important, why would someone pay hundreds to thousands of dollars to get information that may or may not be correct, may or may not be interpreted appropriately or in context, and may or may not be appropriate, applicable, or even necessary? Since most people do not realize the complexity of genomic issues, and are trusting in nature, it is not hard to see how these companies could easily exploit them.

Despite all of this, there is one time that I have agreed with DTC genetic testing. I am a fan of the TV show "Forensic Files." In one episode, a wife used DTC genetic testing to free her wrongfully convicted husband and help catch the real perpetrator. She followed several suspected perpetrators around, obtained DNA samples from discarded items such as cups in fast-food restaurants, sent them away for testing, shared the test results with the detectives, and had her husband's conviction overturned (4).

Now that's good science.

References

1. Sabel MS, Strecher VJ, Schwartz JL, Wang TS, Karimipour DJ, Orringer JS, et al. Patterns of internet use and impact on patients with melanoma. J Am Acad Dermatol 2005;52:779–85.
2. Today. "I'm passionate about life"—actor Tom Cruise talks with "Today" host Matt Lauer about his new love, new movie, and his recent controversial comments. http://www.msnbc.msn.com/id/8343367 (Accessed March 27, 2008).
3. Wohlsen M. Home bipolar disorder test causes stirs. http://www.comcast.net/news/articles/health/2008/03/22/Bipolar.Gene.Test/ (Accessed March 26, 2008).
4. Forensic Files. No witnesses, no leads, no problem. http://www.forensicfiles.com/episodes.htm (Accessed April 7, 2008).

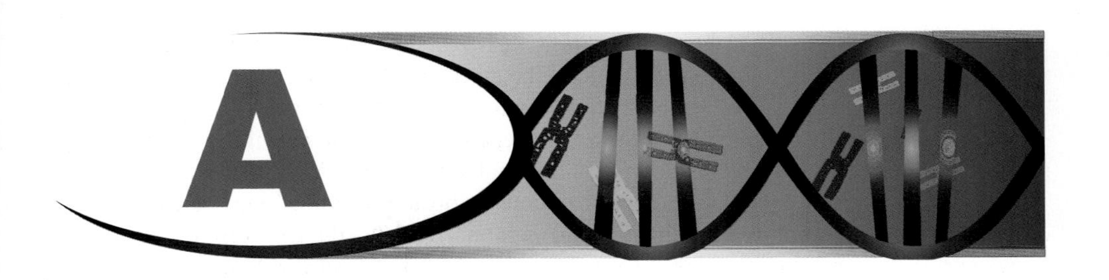

AACC, AMP, and Clinical Laboratory Medicine

The American Association for Clinical Chemistry (www.aacc.org) and the Association for Molecular Pathology (www.amp.org) are examples of professional societies made up of those who work in clinical laboatory medicine, notably, chemists, pathologists, and molecular diagnosticians. AACC is the premier clinical laboratory organization for many fields in laboratory medicine, of which molecular diagnostics is one; AMP is more specialized ("all molecular, all the time").

So what's clinical laboratory medicine? The clinical laboratory, as distinct from the research and discovery laboratory, is where your tumor biopsy, throat swab, urine specimen, or blood sample is sent for analysis by professionals so that specific testing can be performed and information generated that aids your physician in diagnosis and disease management. Some call the pathologist, and by extension the PhD-level laboratory professionals and medical technology professionals who work in the clinical laboratory, the "doctor's doctors."

You may be familiar with the notion of measuring, for example, glucose levels and red blood cell counts to answer questions about diabetes and anemia, respectively. In the last two decades, we have learned how and where specifically (within the long double helix) to examine DNA (and its sister molecule, RNA) to answer all manner of clinical questions. That we can answer questions about genetic diseases using the genetic material is likely not surprising. We have also learned how to use DNA to answer questions about cancer, infectious diseases, compatibility for organ transplantation, metabolism of certain medications, and much more.

New applications arise regularly and are often trumpeted in the lay press. Before you ask your physician about a test you may have heard about, consider its status (research or clinical), whether the test is FDA-cleared or laboratory developed (both are generally fine), and if the testing is done in an appropriately regulated clinical laboratory (qualified labs are certified by the Federal government under CLIA, and by state governments and/or by accreditation bodies like the College of American Pathologists; look for these "Good Housekeeping"-like seals of approval).

A great place to learn more about laboratory testing, DNA-based or otherwise, and its applications to your health, is AACC's highly decorated, highly recognized, highly readable website www.labtestsonline.org. The more applied side of being a scientist or clinical laboratory professional can be appreciated by scanning AACC's and AMP's websites. These careers can be very rewarding and fulfilling.

A-DNA

Companies call themselves AAA, Inc., or AAAAA Widgets, Inc., so they can get top billing in the phone book and enhance their chances at business. Well, why would you think

DNA is any different? (The phonebook—now there's an anachronism in this Web search-dominated world; if we've forced you to go to the dictionary to look up "anachronism" then we've already begun to fulfill our educational mission.)

DNA is naturally organized as a long, double-helix-shaped, string-like structure inside the cells of our bodies. This was described for the first time in 1953 by Francis Crick and James Watson at the Cavendish Laboratory about an hour outside of London; to celebrate their discovery, they went to a nearby pub and announced they had found the "secret of life." Crick and Watson intuited that DNA is made up of two strands wound around each other in a right-handed coil. The strands are made up of chemical compounds called nucleotide bases (ring-shaped structures composed of basic elements: nitrogen, oxygen, phosphorus, carbon, and hydrogen). The nucleotide bases bind to each other on opposite strands of the helix in a defined way [☞ "Complementary strands of DNA"].

The natural way in which the nucleotide bases bind generates the form of DNA ordinarily found in living cells, called the B form of DNA, or B-DNA. The way the bases bind to each other can be changed subtly, under laboratory conditions, so that unusual shapes, angles of binding, and distances in the DNA molecule occur within the DNA double helix; this form is known as A-DNA. Dr. Rosalind Franklin (working at London's King's College in 1952) took some very sharp X-ray-based photos of B-DNA, one in particular (the now-famous photo # 51; see page 66) that led to Watson's and Crick's unraveling of B-DNA's double helical nature. The scientific competition around the race to unravel DNA's mysteries is well chronicled in the books listed in the section of this book called "Further Reading." These books also provide insight into the professional lives of scientists [☞ "B-DNA"].

Agarose

Similar to Jell-O® but much less tasty (actually we never taste the stuff; OSHA, the Federal Occupational Safety and Health Administration, would issue lots of fines to our host institutions). Somewhere between the consistency of Jell-O® Jigglers and a Jell-O® mold is where you would find agarose gels. In the lab, powdered agarose (derived from seaweed) is mixed with water and some salts and microwaved until it boils (there's a lot of kitchen-style procedure in the clinical laboratory; generally if you're good in the kitchen, you're good in the lab, and vice versa). The mixture is cooled and then the liquefied agarose solution is poured into a Jell-O®-type mold called an electrophoresis chamber [☞ "Electrophoresis"]. A comb that looks like this—┬┬┬┬┬┬—is placed into the liquefied agarose, which, as it cools, hardens around the teeth of the comb. The comb is removed, revealing the indentations or wells in the agarose that formed where the teeth were. Now a patient's purified DNA solution may be added to these wells. Electrophoretic analysis follows (the six teeth of the comb form six wells that then allow analysis of up to six DNA samples) [☞ "Electrophoresis"].

This type of electrophoresis is still in use but is increasingly being replaced by more miniaturized, chip-based, and capillary-based technologies, both of which promote automation through microfluidics [☞ "DNA sequencing"]. A lot of engineering is part of today's clinical laboratory medicine field, which is about a $30 billion industry worldwide.

Allele

An allele is a copy of a gene. Genes exist in potentially different forms and are said to exist as alleles (or forms) of that gene. But let's start at the beginning—the human body.

All somatic cells in the human body are diploid, i.e., they have two full "sets" of DNA-containing chromosomes. "Somatic" in fact derives from a Greek word, *somatikos*, meaning "bodily." In this case somatic cells are distinct from the human germ or sex cells called gametes. There are somatic cell mutations, which are not heritable, and germ cell mutations, which are heritable. The two classes of mutations have very different ramifications vis-à-vis the potential need for increased regulatory oversight of genetic testing, generally a controversial topic.

Each somatic cell contains two sets of 23 DNA-containing chromosomes, comprising the 46 chromosomes found in diploid human somatic cells. There are exceptions. No DNA is present in mature red blood cells. Gametes (male sperm and female eggs) have one set of chromosomes and are termed haploid. Sometimes tumor cells, being the ornery, unpredictable, unwelcome critters that they are, don't have the normal complement of chromosomes; they can be diploid or triploid (three sets of chromosomes) or aneuploid (some unusual combination not necessarily divisible by 23).

Let's get back to all the other cells that are actually diploid. A normal diploid cell has two "doses" of each gene, one on each of the two chromosomes present. For example, let's use the gene for the protein that, when expressed, gives rise to an individual's eye color. A person may have two copies of the brown eye color gene, two of the blue eye color gene, or one of each. Those two copies of the gene are the alleles of the gene. Genes exist in potentially different forms on the two chromosomes present and are said to exist as alleles (or forms) of that gene. Someone with two alleles of the same gene is said to be homozygous for the presence of that gene. Getting back to our example, someone with the brown and the blue allele is said to be heterozygous for the presence of that allele; this person has one of each allele. It's the expression of these genes that give rise to proteins that make our irises blue or brown, for example.

For many genes, such as eye color, these differences result in individual variation; a "right/wrong," "good/bad," "normal/abnormal," or "wild-type/mutant" designation is not used. In some instances, however, a "bad" gene can result in an abnormal or mutant protein. This may be the result of a single mutation or different mutations. The designation homozygous wild type is used when both alleles contain the gene that codes for the normal or wild-type protein. If one allele contains a gene that is normal and the other allele contains a gene that is mutant (coming from one parent), this is called heterozygous (as in the above example). When both alleles contain copies of the mutant gene, this is known as a homozygous mutant genotype [☞ "Genotype"and "Phenotype"].

Allele-specific primer extension (ASPE)

An allele is a copy of a gene, and a gene may exist in more than one form. Allele-specific primer extension is a laboratory method used to amplify (so that we have plenty to analyze) one of these forms or a specific allele of interest.

Let's say we would like to know if an individual has inherited a specific gene from a parent who happens to have a history of blood clots. We know of one gene that contains a point mutation (single nucleotide change), which results in a protein that may increase risk for dangerous blood clots. In this case, we know that a normal allele contains the nucleotide adenine (A) and the mutant allele contains the nucleotide guanine (G) at a specific position; think of these as simple spelling differences. Since we know the position of the mutation (it was discovered in the mid '90s), we can use this knowledge to apply a technique to identify the mutant allele (the spelling error).

TECHIE: In ASPE, two oligonucleotide primers that differ at the terminal 3'-nucleotide (in the above example, the A and G nucleotides) are employed. One primer has a thymine (T), which is complementary to A in the wild-type allele, and the other primer has a cytosine (C), complementary to G in the mutant allele. Upon extension of the primers by an enzyme, a laboratory "tool," only the primer with an exact match on the 3'-end will be extended. Identification of the extended product is accomplished by any one of a number of methods, e.g., labeling the two primers with differently colored fluorescent molecules, immobilizing the primers to a solid support such as a different colored bead (www.luminexcorp.com), or using melting curve analysis.

Amplicon

"Amplicons" are the DNA copies made during an amplification method such as PCR or TMA [☞ "PCR" and "Transcription-mediated amplification"]. For amplicons made during the PCR reaction, it is the sequence generated between and including the two primer sequences. Put another way, if the DNA double helix is a book and the target of interest within the DNA is a page in that book, then amplicons are the photocopies made in what is essentially a photocopying or amplification event.

TECHIE: Once amplicons are made, they can be identified using several different methods. (More or less every DNA-based test in the clinical laboratory consists of three parts: (1) extraction of the DNA, (2) amplification of a specific target of interest within that DNA, and (3) identification/detection of that specific target.) Gel electrophoresis is a simple, routine method used for identification and for assessment of the size (molecular weight) of the amplified target. This method alone will provide amplicon size, which may sometimes be enough to identify it. Another level of quality control may be provided by using a "probe" to further characterize the target for correct DNA sequence; the resultant probe/target "hybrid" may be detected by light- or fluorescence-generating techniques [☞ "Probe"].

Amplification refractory mutation system (ARMS)

TECHIE: This is a really fancy, yet descriptive, name for a simple type of PCR reaction that can distinguish two sequences that differ by a single nucleotide. The reaction takes place in two separate PCR tubes using a total of three primers: two forward and one reverse. The two forward primers are identical except at their 3' terminal residues (ends), which are specific to one of two possible sequences, i.e., wild-type or mutant. The reverse primer is complementary to both sequences. Since an exact match at the 3' residue is required for extension by *Taq* polymerase, only the tube containing this exact match to the primer will result in a PCR product, which can be made visible on an agarose gel [☞ "PCR" and "Allele-specific primer extension"].

Analyte

We always get a kick out of it when the definition for a word uses a similar word form in the definition, e.g., "Dissection: The act of dissecting something for scientific or medical study." This is often the case when looking up the word analyte. Most often an analyte is defined as a substance that is undergoing analysis or being analyzed. In clinical laboratories, the analyte may be a chemical entity like glucose or calcium; a cellular component like white blood cells, platelets, or cancer cells; or a foreign substance like a drug or virus. In molecular testing, the analyte is DNA or RNA, which may be from the patient's cells, external donor cells or may be the result of a viral, bacterial, or fungal infection.

Analyte-specific reagents (ASRs)

Analyte-specific reagents (ASRs) are individual components or "building blocks" that are used in laboratory-developed tests (LDTs) [☞ "Laboratory-developed tests"]. An example: a primer set used to identify a single gene or gene segment; by definition ASRs can only be used to identify a *single* analyte. This primer set would be stand-alone and not associated with other components necessary to complete the assay, i.e., enzyme, buffer, dNTPs. Since ASRs are individual reagents and not meant to be commercially associated with a specific test or procedure, their manufacturers are not allowed to promote them as "kits" or claim performance characteristics. The FDA regulates manufacturing and usage of these reagents (http://www.fda.gov/cdrh/oivd/guidance/1590.html).

Anneal

DNA, normally double-stranded in nature, can be manipulated in the laboratory in a number of ways to make it single-stranded. Making double-stranded DNA single-stranded (a process known as denaturation) can be done by heating it to near-boiling temperatures or treating it with strongly alkaline solutions. This process is performed when we want to ask a laboratory question about a particular DNA preparation, for example, "Is an infectious organism's DNA present in the mixture?" or "Does this patient's DNA have a particular genetic mutation?"

One way we answer such questions in the molecular pathology lab is by using a small piece of DNA (known as a probe) that has complementarity to the target of interest, e.g., the microorganism or mutation [☞ "Complementary strands of DNA;" "Denature;" "Probe"]. In other words, the probe has the right matching sequence to seek out and find the target DNA sequence. That process of binding a probe to a target is a joining of two pieces of complementary DNA in a biochemical process known as "annealing." Think of probes as highly specific pieces of "molecular Velcro™."

In the example cited, a special form of annealing (called hybridization) has occurred, because a hybrid DNA duplex between synthetic probe and natural bacterial or human DNA has been formed. During PCR, short stretches of nucleotide chains (oligonucleotides) called primers anneal to complementary DNA sequences in the target as part of the overall performance of those reactions [☞ "PCR;" "Primer;" "Real-time PCR"].

Antibiotic resistance

Penicillin was discovered in 1928 by Alexander Fleming in London. Many new antibiotic discoveries and formulations have followed in the decades since. In the last decade or so we have found that significant percentages of hospital isolates of *Streptococcus pneumoniae* and *Staphylococcus aureus* are partially or completely resistant to penicillin and methicillin, respectively. (By convention, living organisms are named by binomial nomenclature, a fancy term that means biologist's name—nomenclature—and refer to the genus and species of an organism like *Homo sapiens* with two words—binomial. The name is italicized because it is Latin and it is spelled out in full the first time it appears in writing. Subsequent mentions abbreviate the term by simply using the first letter of the genus as in *S. pneumoniae*.)

Disease-causing bacteria, or so-called human pathogens, may acquire resistance to antibiotics by genetic mechanisms. If antibiotics in the environment are shared by human pathogens—for example, when an infected patient is being treated—there may be one or a few bacteria that adapt (by mutating) to biochemically bypass the killing action of the antibiotic. The drug is said to have selected for the resistant bacteria. Or, pathogens may acquire the DNA that codes for antibiotic resistance by cross-species transfer, also-called "trans-species leap." Think of that as sex between bacteria (Figure 1).

The normally harmless species of Enterococcus that lives in the human gut is suspected to have passed its gene for tetracycline resistance to *S. pneumoniae* and *Neisseria gonorrhoeae*. Tetracyline-resistant forms of pneumonia and gonorrhea are the result. There are many other examples. The bottom line is that if antibiotic-resistant bacteria are present, they quickly outgrow the antibiotic-sensitive bacteria in a patient and become a serious clinical issue. The individual and public health "take home" lessons of these phenomena include taking all of your prescribed antibiotics to minimize the chance of disease flare-up from unkilled, mutated bacteria.

It's equally important not to pester your physician for an antibiotic prescription if an illness is virally induced. Antibiotics don't kill viruses, and taking an antibiotic when one has a

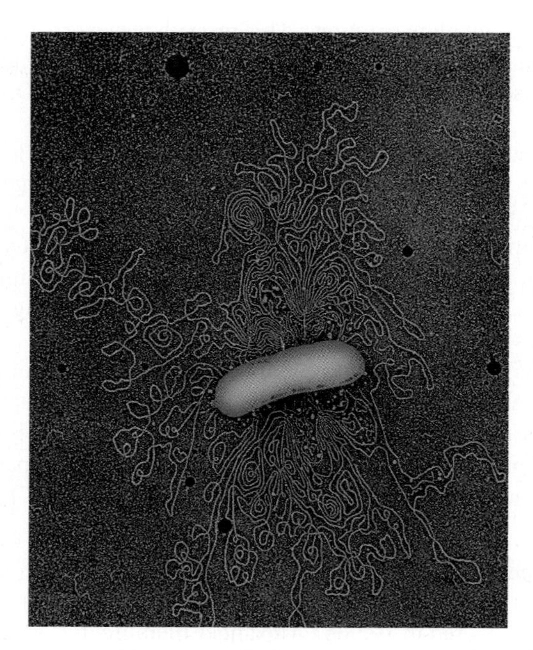

Figure 1. DNA from a ruptured bacterial cell; the large amount of DNA from this ruptured bacterium demonstrates how tightly packed DNA is inside a (bacterial) cell.

viral infection only selects for drug-resistant bacteria, which may harm you. A diagnostic test that can rapidly and cheaply differentiate between a virally induced and bacterially induced illness when you take your child to the pediatrician would fill a need. You can learn more about antibiotic resistance by reading about it in the April 23, 2002, issue of *Scientific American* ("Agricultural Antibiotic Use Could Contribute to Drug Resistance") or by going to the Centers for Disease Control and Prevention web page (http://www.cdc.gov/drugresistance/community/).

Some European countries have successfully reduced the use of antibiotics as a routine additive to pig feed. While antibiotics in feed tend to modestly improve animal growth, they also contribute to the selection of antibiotic-resistant bacteria. This is true not only in animals but potentially in humans who ingest their meat, thereby possibly receiving low-level antibiotic doses. Indeed, Nobel laureate J. Michael Bishop has termed the alarmingly high use of antibiotics "promiscuous." Slow progress is also being made on this front in the U.S. where there are lobbying and political hurdles to overcome. (See http://www.hsus.org/farm/resources/research/pubhealth/human_health_antibiotics.html)

Here are some staggering figures, circa 2003, borrowed from J. Michael Bishop's book *How to Win the Nobel Prize* (Harvard University Press, 2003):

- Seventy percent of antibiotics manufactured in the U.S. are given to healthy livestock.
- While only ~20% of adults in the U.S. with sore throats can benefit from a round of antibiotic therapy, ~75% get a prescription for antibiotics.
- Approximately half of all antibiotic prescriptions written annually in the U.S. are unnecessary.
- More than 90% of pathogenic staphylococci are penicillin resistant; some strains are resistant to every known antibiotic (read those last eight words again).

Anticodon

TECHIE: but really interesting—anticodons are three-nucleotide sequences specific for a target in messenger RNA (mRNA). Let's back up. DNA is transcribed into mRNA, which is translated into proteins [☞ "Central dogma of molecular biology;" "Expression;" "Genetic code"]. The sequence contained in mRNA (which was dictated by the DNA, or gene, that encoded it) is translated by the cellular machinery into proteins that carry on the business of life throughout the body.

Protein synthesis inside the cell is a rather complicated, yet beautiful, biochemical process. In short, it happens inside a protein-synthesizing "machine" called a ribosome. The ribosome is where mRNA and amino acids come together in a specific way, and the protein that is encoded by that mRNA molecule elongates, amino acid by amino acid, until it's done. The amino acids in the cell (they got there because you had a burger for lunch or a glass of milk with lunch) get to the ribosome for protein synthesis because they are taken there by a molecule that commandeers them; that molecule is called transfer RNA (tRNA; see Figure 2) and looks something like a cloverleaf.

There are specific tRNA cloverleafs for specific amino acids. In fact there are more than 20 tRNAs and each is about 75–100 base pairs long. The amino acid binds on "top" of the cloverleaf, and on the bottom is a three-base-pair sequence called an anticodon. Based on the laws of complementarity [☞ "Complementary strands of DNA"], the anticodon binds to the specific sequence in the mRNA, now associated with the ribosome, that happens to

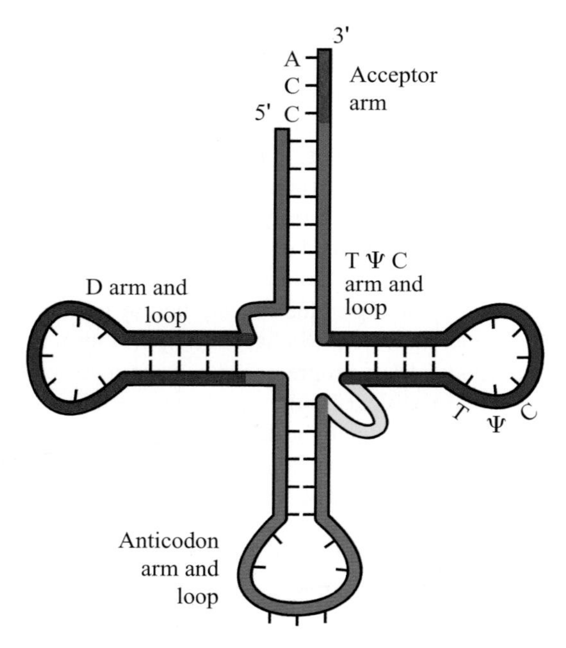

Figure 2. tRNA molecule; specific binding between mRNA and the anticodon portion of the tRNA facilitates donation of the correct amino acid at the top of the tRNA cloverleaf (acceptor arm) during protein synthesis.

Reprinted with permission from Gregory B. Gloor, PhD, Department of Biochemistry, University of Western Ontario, and http://www.biochem.uwo.ca

code for the amino acid leucine (an example). There is a specific codon in mRNA for leucine and a specific anticodon on the bottom of a leucine tRNA cloverleaf that recognizes the sequence, binds there, and gives up the leucine at the top of the tRNA cloverleaf to the growing protein chain in the ribosome. Think of the tRNA cloverleaf as the adapter that translates the nucleotide language of mRNA into the amino acid language of proteins, one "word" at a time.

Antiparallel

The two strands in a DNA double helix are antiparallel to each other (Figure 3). Chemically speaking, each strand or chain is made up of repeating units of deoxyribonucleotides linked one to the next. Deoxyribonucleotides are composed of phosphate groups, a five-sided sugar molecule, and nitrogen-containing bases. Each of the positions in the sugar molecules is numbered, and the phosphate groups serve as chemical bridges attaching the nucleotides one to the next. These phosphate bridges link the 3-position of the sugar in one nucleotide to the 5-position of the next. So that strand runs 3-5-3-5-3-5-, etc. The other strand runs in the opposite direction: 5-3-5-3-5-3-, etc. The two strands are said to be antiparallel to each other due to their chemical structures. Think of this as a two-lane highway, each lane parallel to the other but with traffic flowing in opposite directions in each lane.

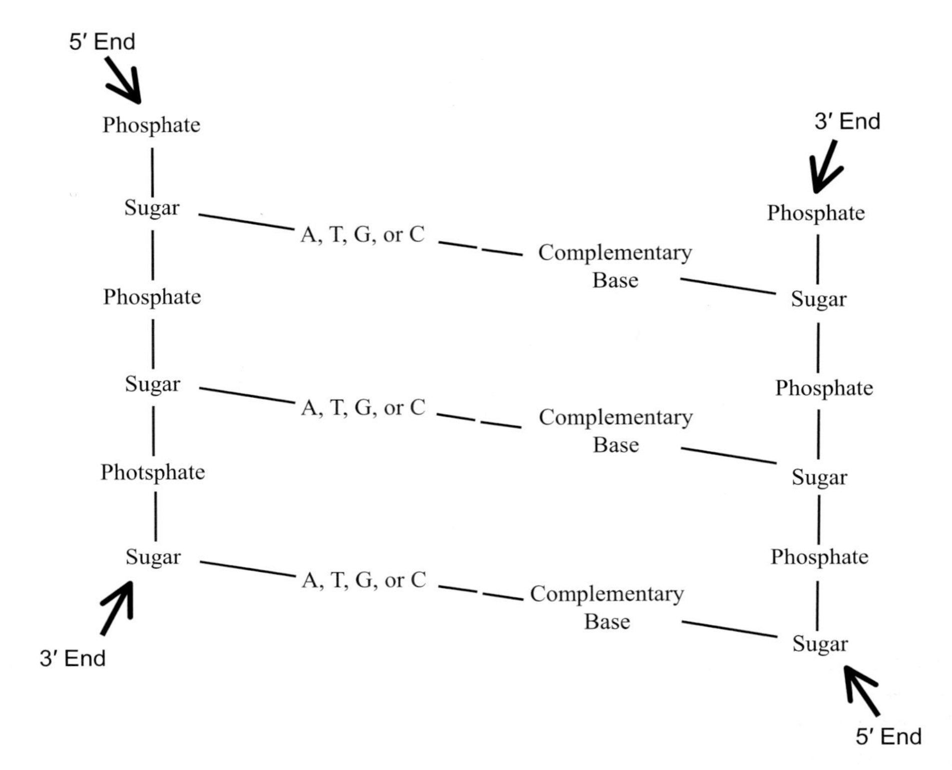

Figure 3. The antiparallel nature of DNA

Apoptosis

(some pronounce the "pop" part but many pronounce the first "o" as a "u" and leave the second "p" silent [a-puh-to-sis])

Programmed cell death. Various genes in the cell encode proteins that are like an internal demolition company. These apoptotic proteins have jobs involving the shattering of the cell's nucleus, cutting up its chromosomes, degrading the internal skeleton or protein-based scaffolding that holds the cell together and defines its shape, and generally destroying the cell and fragmenting it into smaller pieces for disposal. The natural place for apoptosis is during growth, when old cells must give way to the new, or when a cell is infected.

Apoptosis, the act of cellular suicide, is highly regulated and subject to different environmental factors. Too little regulation and the volume control on (appropriate) cell death is turned down and immortalization (cancer) may ensue. If the volume control is turned up, as in stroke or the neurodegenerative condition Parkinson's disease, inappropriate cell death ensues. The proteins that pronounce, initiate, and carry out these cellular death sentences are called caspases.

Armored RNA

RNA is a molecule that is readily and easily degraded, largely due to its structure and the ubiquitous presence of ribonucleases (enzymes that degrade RNA) in the environment

[☞ "Ribonuclease"]. Because of this, maintaining stocks of RNA in the laboratory for use as controls, standards, or calibrators (all useful items in performance of clinical laboratory testing) is difficult. In order to "protect" the RNA from degradation, coat proteins from certain bacteriophages (viruses that infect bacteria) have been employed as an "armor" for RNA molecules. To make armored RNA, the desired RNA sequence is usually produced in a bacterium such as *Escherichia coli* and packaged into the viral particles in place of the normal viral RNA; the resultant "armored RNA" is then resistant to degradation for extended periods. Armored RNA was pioneered by a company called Ambion.

One popular application for armored RNA in clinical laboratory testing is as a positive control for RNA-based assays such as the HIV viral load assay, used in monitoring the viral burden of HIV-infected patients. Armored RNA HIV control reagents contain only a fraction of the entire HIV sequence. Furthermore, the coat proteins of the "armor" possess no receptors for human cells. Thus this example of armored RNA is perfectly safe for use by clinical laboratory personnel.

Array

In the context of molecular diagnostics and laboratory testing, an array can be thought of as a group or panel. For example, in assessing which of the five or ten most common causes of respiratory infection is present in a patient specimen, an array of causative microorganisms, viruses, and/or bacteria, may be assessed simultaneously in an array to learn the specific culprit, which is useful in disease management [☞ "DNA chips"].

"ase"

This is a suffix that denotes an enzyme, a protein that has a specific biochemical job. There are words ending in "ase" throughout the book. Proteases are enzymatic proteins that degrade other proteins (for the most part all enzymes are proteins but not all proteins are enzymes). RNases are enzymes that degrade RNA. Synthases are the general class of enzymes that synthesize other molecules. Polymerases are enzymes that polymerize the formation of other molecules, e.g., DNA polymerase synthesizes new DNA. Capases are enzymes that pronounce, initiate, and carry out cellular death sentences. There are many other examples.

AUG

Yes, it's the abbreviation for the month, August, but with respect to molecular biology, AUG is quite significant. Inherent in the genetic code is information ultimately translated from DNA into protein through the intermediate called mRNA [☞ "Genetic code;" "mRNA;" "Translation"]. The letters in AUG stand for Adenine, Uracil and Guanine, which are bases found in RNA. The genetic code stipulates that AUG codes for the amino acid, methionine. The first amino acid in all proteins is methionine; AUG is the initiator codon [☞ "Codon"] for protein synthesis. Other triplets of interest in the code are UAA, UGA, and UAG: these three are all "STOP codons" and signal the cellular machinery to STOP or terminate protein synthesis. The genetic code was solved by classic experiments of Marshall Nirenberg, Har G. Khorana, and Robert Holley, for which they shared the Nobel Prize in 1968.

Autoradiograph [☞ "Southern blot"]; jargon term: Autorad

It's like an X-ray that you might get for your teeth or a possible bone fracture. An autoradiograph is the end result of any laboratory technique that uses a combination of radioactively labeled components and X-ray film exposure (such as Southern, Northern, Western blots) as well as old-fashioned DNA sequencing gels. The following example is based on one of these techniques, the Southern blot, a technique with continuing limited utility but moving toward obsolescence. Still, the description that follows illustrates many practical aspects of molecular biology.

The goal of a Southern blot is to identify a particular gene or gene fragment buried within the DNA purified from a patient specimen; it is very much akin to looking for a needle in a haystack (actually it's more similar to looking for one piece of hay within the haystack). One way to do this is to cut the patient's DNA into small pieces, separate the pieces using electrophoresis, transfer the DNA to a solid support, and then identify the gene of interest through hybridization using a radioactively labeled probe [☞ "Restriction endonucleases;" "Probe;" "Hybridization"]. The probe serves two functions: (1) to identify the sequence of interest through complementary binding and (2) to enable visualization via the radioactive label "built into" the probe. After the gene fragment of interest and the probe have hybridized, X-ray film is placed on top of the solid support containing the DNA; the radioactive hybrid (target DNA plus radioactive probe) exposes the film. The film is then developed using standard film developing methods, generating the autoradiograph ("auto" because it exposed itself, "radio" for radioactive, and "graph" to mean something like a photograph). The radioactive label increases the sensitivity of the procedure by amplifying the signal from small starting amounts of DNA.

Visual inspection of the autoradiograph allows us to determine if the specific piece of DNA was present, in order to answer questions about the presence of some aspect of that patient's DNA that might be instructive in making a particular diagnosis. In fact, in a nod toward safety, radioactive labels in this technique were largely replaced in the '90s by other methods such as luminescence. A DNA probe that is not radioactive but rather luminesces (generates light) under the right conditions generates in the above-described procedure a film called a "lumigraph." Sometimes the synonym autoradiogram or lumigram is used instead of autoradiograph or lumigraph (Figure 4).

Autosome

Autosomes are all the chromosomes present in a cell except for the two sex chromosomes, X and Y. The human cell consists of 22 pairs of autosomes and one pair of sex chromosomes.

Avery, Oswald T. (1877–1955)

In 1944 at the Rockefeller Institute (now University) Hospital in New York City, Dr. Avery and his colleagues Colin MacLeod and Maclyn McCarty showed, through a series of classic experiments with strains of bacteria that cause pneumonia (*Diplococcus pneumoniae*, now called *Pneumococcus pneumoniae*), that DNA carries genetic information. This work extended the findings of Fred Griffith, an English bacteriologist, who initially made

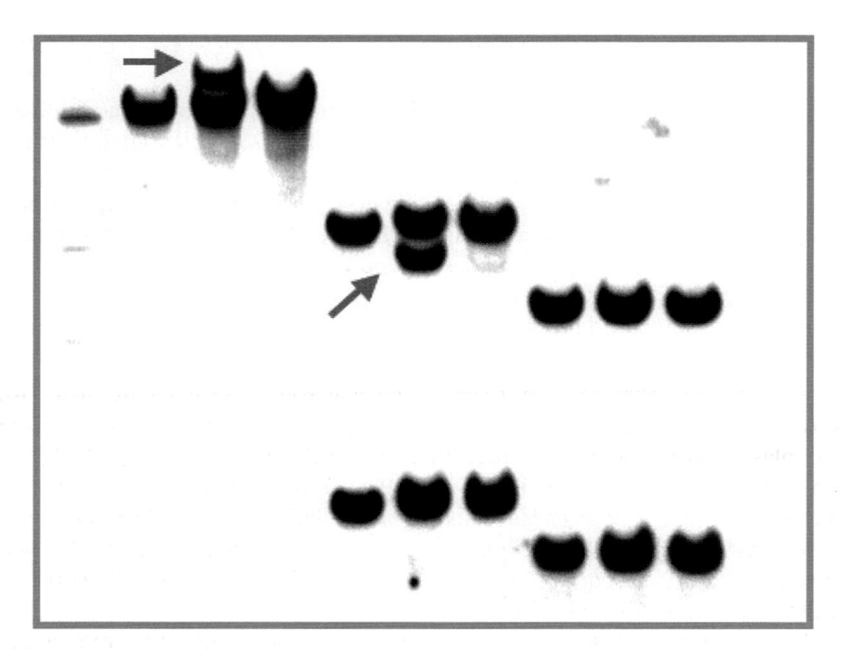

Figure 4. An autoradiograph of DNAs obtained from patients and analyzed to demonstrate the presence or absence of specific DNA rearrangements indicative of leukemia or lymphoma. The arrows point to DNA bands diagnostic for a particular type of leukemia in this patient.

Reprinted with permission from Tsongalis GT, Coleman WB. Molecular diagnostics: a training and study guide. Washington, DC: AACC Press, 2002.

observations in 1928 about the transfer of genetic material. These experiments are considered biological historical landmarks.

The phenotype [☞ "Phenotype"] used as a marker in these experiments was virulence (or ability to kill) in mice. When injected into experimental mice, non-virulent pneumococci did not kill the animals. When non-virulent bacteria were transformed in vitro with DNA from virulent bacteria, the non-virulent bacteria gained the virulence phenotype (and genotype); they became able to kill experimental mice. Furthermore, if the DNA used to transform the non-virulent bacteria was first treated with DNase, an enzyme that specifically degrades DNA, no change occurred, helping to cement the notion that DNA is the responsible agent for the carrier of genetic information.

Avery and his colleagues' landmark work flew in the face of the day's conventional wisdom that proteins were the genetic material. Their finding that DNA was the genetic material was, of course, validated many times over.

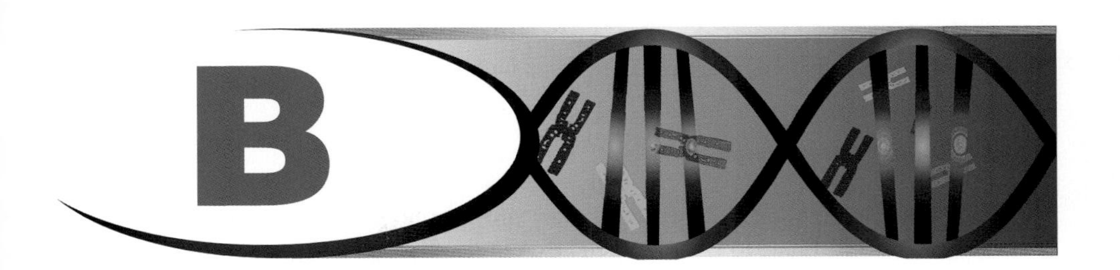

BAC

Bacterial artificial chromosome. To some, this may sound like it belongs in a "B-rated" science fiction movie with huge bacteria slithering around destroying mankind. In reality, BACs are plasmids (laboratory constructs that act as vehicles for carrying DNA) that can accept large inserts of DNA for cloning, generally in *E. coli* bacteria [☞ "Plasmid"]. The difference between BAC plasmids and "typical" plasmids are not only the larger size of the DNA insert that BACs can accept, but also that BACs are based on a fertility plasmid (F-plasmid), which ensures equal distribution among dividing bacterial cells. This is important for obtaining large quantities for subsequent use or testing.

BACs are useful in DNA sequencing projects and were used in the Human Genome Project. For example, human DNA fragments may be inserted into BAC plasmids to determine their sequences—and then the sequences from many, many BAC plasmids are pieced together to make a complete sequence [☞ "Human Genome Project;" "HAC;" "YAC"].

Bacteriophage

Phage is another word for virus and the entomology is once again Greek (phagein, meaning "to eat"). Viruses don't just prey on humans. Certain viruses specifically infect different kinds of bacteria. Examples of bacteriophage include lambda (λ), T4 (Figure 5), and Qβ.

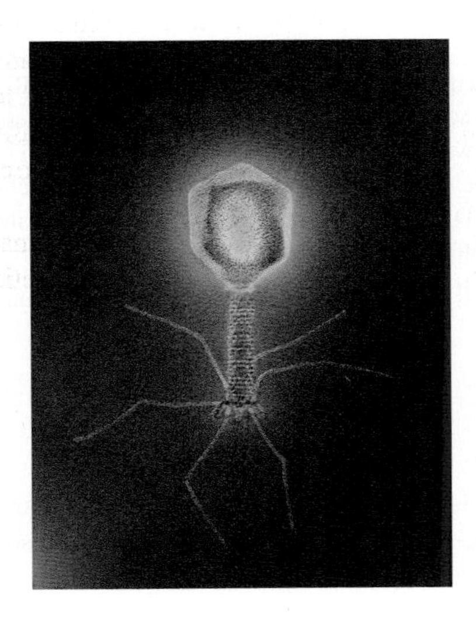

Figure 5. T4 bacteriophage. Colored transmission electron micrograph (TEM) of a T4 bacteriophage virus. The swollen structure at top is the head, which contains DNA inside a protein coat. Attached to this is the tail, consisting of a tube-like sheath and tail fibers (at bottom). T4 bacteriophages are parasites of *Escherichia coli*, a bacteria common in the human gut. The virus attaches itself to the host bacteria cell wall by its tail fibers; the sheath then contracts, injecting the contents of the head (DNA) into the host. The viral DNA makes the bacteria manufacture more copies of the virus.

Scientists have learned much about gene expression, in general, by studying the simple genomes of these organisms and their life cycles within their bacterial hosts [☞ "Virus"].

Band [☞ "Electrophoresis"]

There were some great ones when we were growing up in the '60s and '70s. But how does the term "band" relate to DNA? When DNA is electrophoresed (physically moved and separated based on electrical charge) in order to study it, it is placed into an electrophoretic gel well that is shaped like a rectangle. As electrophoresis proceeds to completion, DNA fragments are separated and ultimately can be visualized [☞ "Ethidium bromide"]; they retain the basic rectangular shape of the well but have been condensed by the process of electrophoresis into a tight line of visible DNA referred to in the field as a band (Figure 6). The same band-shaped images are generated by the detection phase of Southern blotting [☞ "Autoradiograph" and "Southern blot"].

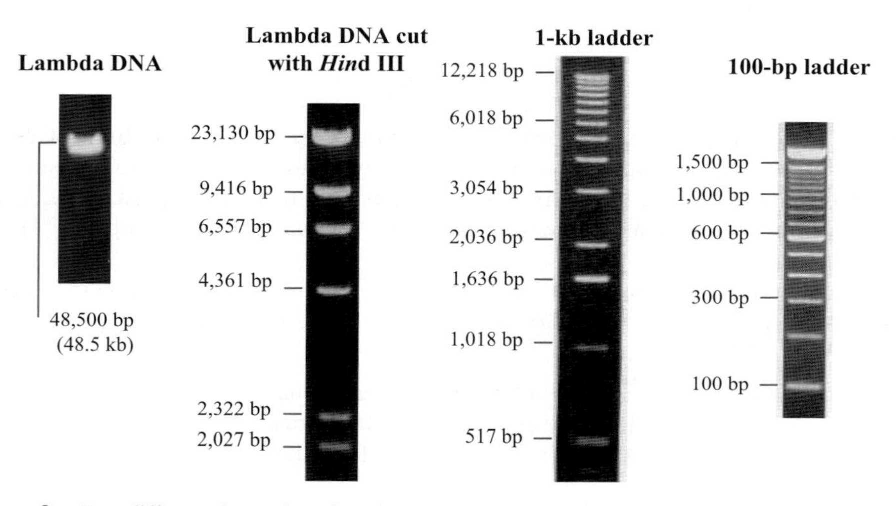

Figure 6. Four different electrophoretic gel segments (unrelated to each other) are shown: lane 1, DNA from a bacteriophage called lambda; lane 2, lambda DNA "cut" into pieces with a protein called *Hin* dIII; and lanes 3 and 4, commercially prepared DNA mixtures with evenly spaced markers, called ladders. Each gel was loaded with a different DNA mixture that was then subjected to an electric field. The differently sized DNA molecules separated based on their molecular weights in the electric field. Heavier pieces migrated more slowly than lighter pieces of DNA. The sizes are labeled in base pairs, and the pieces of DNA, visible because they were stained with ethidium bromide, are called bands.

Reprinted with permission from Tsongalis GT, Coleman WB. Molecular diagnostics: a training and study guide. Washington, DC: AACC Press, 2002.

Bases

Much as we love it, we're not talking baseball—we're talking freshman biochemistry. DNA and RNA are made up of bases, which are ring-shaped chemical structures composed of the elements carbon, hydrogen, nitrogen, and oxygen in various combinations (Figure 7).

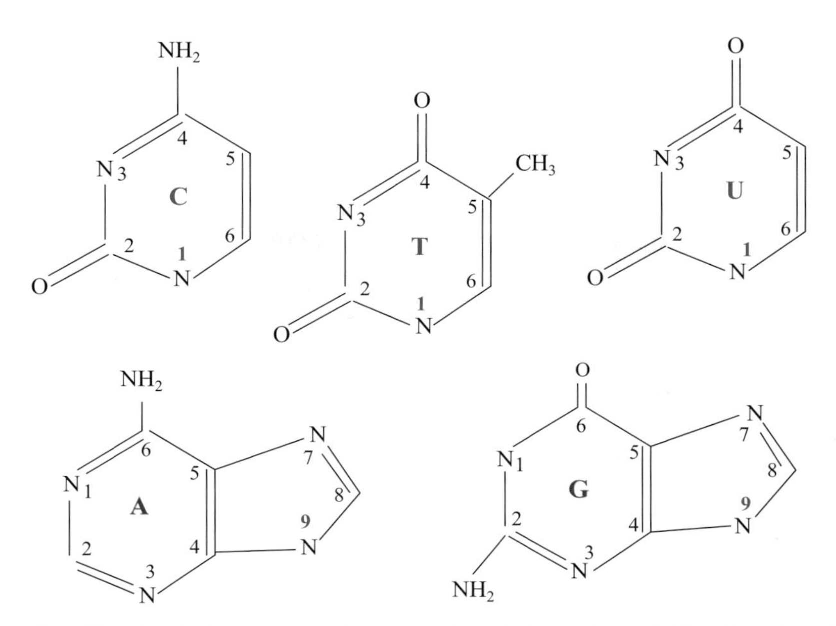

Figure 7. The chemical structures (nitrogenous bases) that make up DNA (C, T, A, and G) and RNA (C, T, U, and G).

Reprinted with permission from Tsongalis GT, Coleman WB. Molecular diagnostics: a training and study guide. Washington, DC: AACC Press, 2002.

DNA is made up of the bases adenine, guanine, thymine, and cytosine (A, G, T, and C, respectively). The bases in RNA have an extra oxygen molecule (hence, ribo instead of deoxyribonucleic acid; the "de" meaning without oxygen) and include A, G, C, and uracil (U); there are no thymine bases in RNA [☞ "Nucleotide" and "Nucleoside"].

B-DNA [☞ "A-DNA"]

The naturally occurring form of DNA inside cells but so-named because it was observed and described after A-DNA. B-DNA has the normal shape and normal angles of binding for the nucleotides that form DNA, and normal distances within the double helix that are found within DNA in the body (Figure 8). The structure of this form of DNA was solved by Watson and Crick. B-DNA was the form of DNA that was famously photographed (using X-ray diffraction) by Rosalind Franklin [☞ "A-DNA" and "Franklin, Rosalind"].

bDNA [☞ "PCR wannabes"]

bDNA is the abbreviation for branched DNA. The basis of a technology competitive to PCR, bDNA is an in vitro nucleic acid amplification technique where the signal used to identify the DNA or RNA target is amplified, as opposed to amplifying the DNA or RNA target itself. So it is more appropriately thought of as a signal amplification technique. A fair amount of biochemistry and enzymology occurs; if the target is present in the patient sample (the target is usually a human pathogen like hepatitis C virus or human immunodeficiency virus),

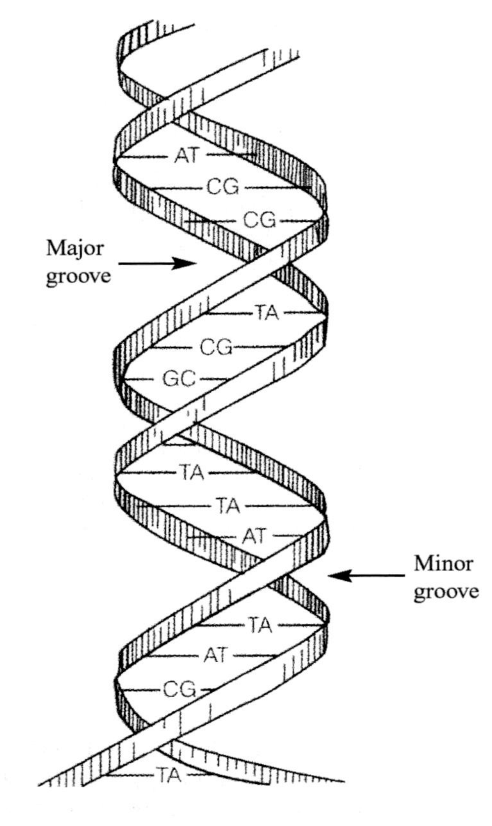

Figure 8. The DNA double helix is held together by hydrogen bonds between adenine-thymine (AT) and guanine-cytosine (GC) base pairs. B-DNA is the biologically important form of DNA.

then a reaction that generates detectable quantities of light occurs and the test is positive. If the virus was not in the initial patient specimen, no light is generated and the test is read as negative by the clinical laboratory performing the test.

Beacons

Generally speaking, a beacon is a signaling device. This is precisely the right term for the oligonucleotide probes referred to as molecular beacons, because they are used to signal the presence of amplified target nucleic acids, the predominant object of investigation in molecular diagnostics tests.

Molecular beacons are oligonucleotides shaped like a hairpin (Figure 9). The loop portion of the hairpin is complementary to a portion of the amplified target nucleic acid of interest. The stem portion of the hairpin is designed so that one half is complementary to the other, thereby causing the hairpin structure. On the ends of the stem are a fluorescent dye and a non-fluorescent quencher molecule that prevents the dye from fluorescing. In the absence of target amplicon, there is nothing to induce the hairpin to straighten outward, since the stem is a stable nucleic acid hybrid, and there is no fluorescence due to the proximity of the quencher to the fluorescent dye. In the presence of specific amplified target, however, hybridization between the loop portion of the hairpin and the amplicon is favored. During that hybridization event, the molecular beacon becomes linear (see Figure 9), the

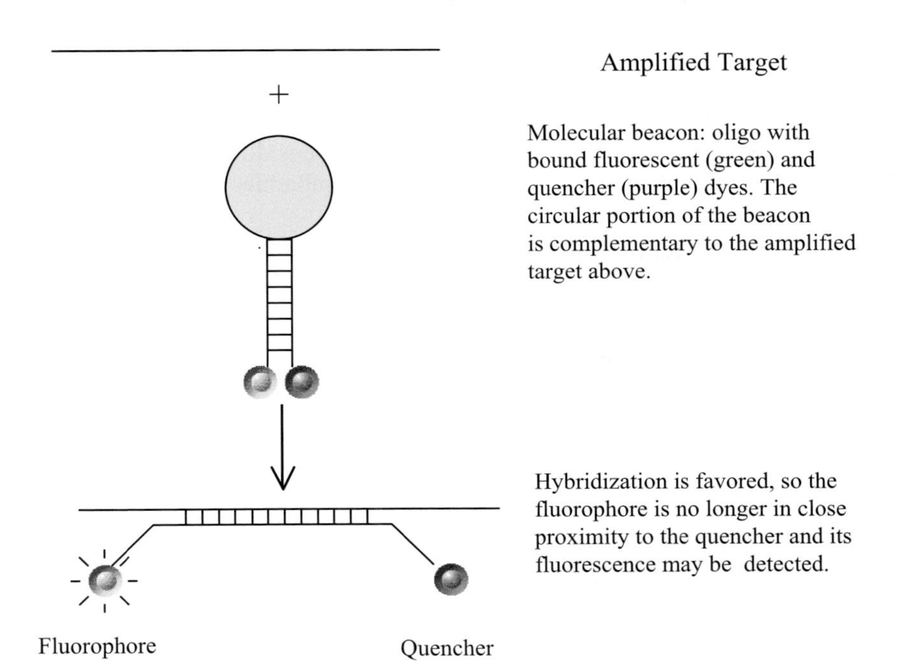

Figure 9.

fluorescent dye is no longer in close enough proximity to the quencher for the quencher to act, and so signal may be detected. (This is accomplished after laser energy is applied and optical detection equipment captures the signal, in any one of a number of commercially available laboratory instruments.)

Molecular beacons, therefore, can be used to report the presence of specific nucleic acids in "real-time" amplification assays [☞ "Hairpins;" "Oligonucleotide;" "Real-time PCR"].

Boxer

You might remember Boxer as George Orwell's symbol for the overworked and exploited masses of the Russian Revolution in his classic book, *Animal Farm*. (Homework assignment: Read *Animal Farm*; remember our goal is to educate and that's not limited to science. Let's "fatten up" common knowledge, which scientific luminaries like Robert Oppenheimer and Michael Bishop have suggested is "thinning.")

In the context of DNA technology, however, the Boxer was chosen as the dog to be used in the sequencing of the canine genome by the NHGRI (National Human Genome Research Institute, a branch of the Federal National Institutes of Health). Specifically, a female Boxer named Tasha was chosen as *the* pooch; Tasha wasn't harmed, of course, because there's plenty of DNA available from any one of a number of her cellular specimens: blood, cheek scrapings, even the material in what Tasha leaves in the pooper scooper. The genetic map of *Canis lupus familiaris* is now complete.

Why sequence the canine genome? Just as we can learn more about basic biology and apply those lessons to understanding human disease by sequencing simple organisms, we can learn even more from sequencing animals closer to humans on the evolutionary scale. In other words, there may be instances where canine biology is closer to human biology than, for example, the mouse or a particular bacterium. Knowledge of both the human and canine genomes (80% the length of the human genome) allows improved understanding for both human and canine disease by comparing similarities and differences between the two genomes.

BRCA1

The *BRCA1* gene was cloned [☞ "Cloned"] in 1994. The *BRCA1* gene, when mutated, is responsible for a fraction of heritable breast cancer (as opposed to sporadic breast cancer, the overwhelmingly dominant form of the disease). There are particular populations in which *BRCA1* mutations are prevalent and mutation screening may be appropriate. But how to proceed?

Individuals with a family history of breast cancer may ask physicians for this test. If the patient is shown by *BRCA1* testing to harbor the same mutation as an affected primary relative, then there is reason for concern; the patient has an increased *lifetime* risk for breast cancer. But the disease may strike at age 30 or age 80. What then?

A drastic but potentially life-saving option is prophylactic double mastectomy. Furthermore, *BRCA1* mutations lead to associated increased risk for ovarian cancer, raising the question of further prophylactic surgery. Even surgery may leave behind some cells that harbor the mutation. The patient and physician may decide that this patient should be subjected to more frequent mammograms. It is possible that discovering a *BRCA1* mutation may result in diligence about mammography for the rest of a woman's life in the same way that we now take for granted the importance of closely monitoring cholesterol and lipoprotein (HDL and LDL) levels in those found to be at risk for heart disease. Such an eventuality, should it come to pass, would be a good thing for the overall public health; all, including insurance companies, should come to realize these possibilities in time. In any case, it is important to consider all therapeutic and diagnostic options in case a *BRCA1* mutation is uncovered. It is equally important to consider these options in the context of professional and appropriate genetic counseling.

On the other hand, such an individual (with a relevant family history) may be tested and shown not to harbor that specific mutation present in the family. In this scenario, she *may* proceed through life with a false sense of security, bypassing regular mammograms and ignoring dietary concerns. Inherited breast cancer represents only a small fraction of all cases of breast cancer. Moreover, *BRCA1* is only one of the genes that may be related to increased risk of breast cancer (*BRCA2* is one of several others). There are hundreds of mutations within *BRCA1* that are associated with breast cancer and testing for one or a few and not finding them is no guarantee that breast cancer will not strike the individual.

The cloning of *BRCA1* was a great scientific achievement. It has generated exciting and useful clinical options. Testing options may be accessed at the website of the company that owns the rights to *BRCA1*, Myriad Genetics (www.myriad.com), which cloned the gene.

For a list of web-based links to information about breast cancer, see http://research.nhgri.nih.gov/bic/resources.shtml.

For more information on breast health/breast cancer, see Table 1.

Table 1. Resources: Breast Health/Breast Cancer

Resource	Phone
National Cancer Institute's Cancer Information Service http://www.nci.nih.gov/cancer_information/cancer_type/breast/	800.4.CANCER
The American Cancer Society http://www.cancer.org/docroot/lrn/lrn_0.asp	800.ACS.2345
The Y-ME National Breast Cancer Organization http://www.y-me.org/	800.221.2141
National Alliance of Breast Cancer Organizations http://www.nabco.org/	888.80.NABCO

Parenthetically, we'll use this entry to make a semantic point. We all have the *BRCA1* gene—male and female alike. Only the mutated form is the one associated with familial breast cancer. People often speak of "having the gene" for disease x, y, or z. What they really mean to say is "having the *mutation* in the gene"—that's what's actually responsible for developing the disease.

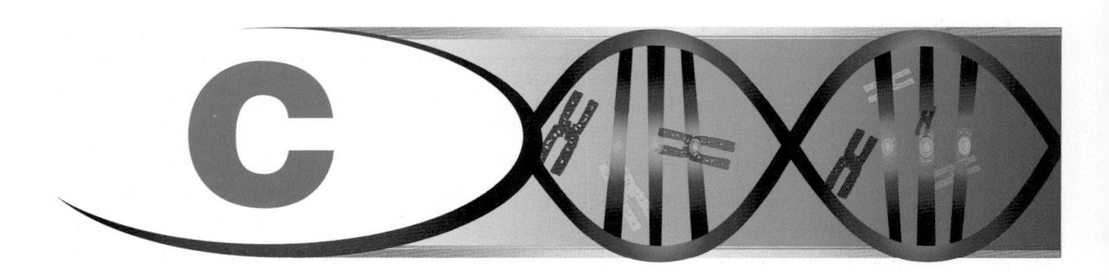

Caenorhabditis elegans

This organism is also referred to as *C. elegans*, which is the convention for naming organisms when writing about them. After the genus (*Caenorhabditis*) first appears in writing, it can then be abbreviated with only the first letter followed by the species name, in this case *elegans*.

C. elegans is a tiny, free-living, transparent worm that lives in the soil and ingests bacteria. This worm is multicellular. There is an entry for this worm in this book because after eight years, in December 1998, *C. elegans* became the first multicellular organism to have its complete genome sequenced—all 20,000 of its genes (humans have only a few thousand more genes than this worm; how's that for promoting humility?).

The *C. elegans* genome contains about 100 million bases, compared to 300 million for humans; clearly *C. elegans* packs a lot more genetic information into a smaller genome, relatively speaking, than the human genome. Scientists have been studying the biology of *C. elegans* for many years. Much is known about the worm's biology and it has great utility as an experimental model system. Knowledge of its genome and experimental manipulation provide insight into basic gene function, which leads to better understanding of human gene function, biology, and disease.

Caspases [☞ "Apoptosis"]

Caspases are proteins involved in programmed cell death (apoptosis). These proteins may be thought of as the judge, jury, and executioner of the cell, with the added twist that they are internal to the cell. It's a good thing sufficient regulation exists to keep caspases and apoptosis in check. Left unchecked, degeneration like that seen in Parkinson's disease can occur. When cells escape natural death, they become immortal and develop into cancers.

cDNA [☞ "PCR"]

Complementary DNA, which is an unusual biochemical entity (sounds like StarTrek technobabble, doesn't it?). DNA is the target of a powerful laboratory method called Polymerase Chain Reaction (PCR).

Sometimes, however, we need to ask a question about RNA, not DNA. Examples include when we are searching in the diagnostics laboratory for the presence of a virus that naturally only contains RNA (for example, hepatitis C virus or human immunodeficiency virus) or if we are investigating not the presence of a gene (DNA) associated with disease but rather the expression [☞ "Expression"] of that gene as RNA.

In such cases we still want to exploit the power of PCR but we first need to purify the RNA of interest and turn it into DNA so that we can proceed with the PCR method in the laboratory. We do that with an enzyme called reverse transcriptase [☞ "Retroviruses" and "Reverse transcriptase"]. We mix the purified patient RNA (which would include an infected patient's viral RNA if that was in fact the target) with Reverse Transcriptase (RT) and other necessary biochemical ingredients. When this enzyme (RT) encounters RNA, it does its job: it turns that RNA molecule into a DNA "copy" of that RNA. That DNA is complementary [☞ "Complementary strands of DNA"] to the RNA that the enzyme used as a template and is called complementary DNA, or cDNA for short. cDNA can then participate in PCR just like any other DNA molecule.

In brief then: RNA + RT = cDNA, which is amenable to amplification by PCR and subsequent clinical laboratory analysis. (*Tth* polymerase is an enzyme with dual properties. It combines the activities of reverse transcription and the important enzyme in PCR, DNA polymerase, whose job is to make more DNA. So it's a handy enzyme to use in RT-PCR, i.e., reverse transcriptase PCR, which allows PCR using RNA, not DNA, as starting material. *Tth* polymerase is an enzyme from the bacteria, *Thermus thermophilus*, hence the name.)

Central dogma of molecular biology

Shortly after the discovery of the structure of DNA, the flow of genetic information from DNA to RNA to protein was described. This process was first described by Francis Crick in 1958 and is known as the central dogma of molecular biology. It is the paradigm on which the field of molecular biology is based.

As we have come to learn, all of a cell's genetic information is contained in its DNA. This genetic information serves as a template to be copied and distributed to daughter cells in a process called DNA replication. Next, DNA is transcribed into mRNA, which is then translated by the ribosome into protein. The central dogma is based on DNA as the starting molecule, with the flow of genetic information always going in one direction: DNA → RNA → Protein. Until the discovery of the enzyme reverse transcriptase in retroviruses in 1970 (independently by Howard Temin and David Baltimore, who jointly won the Nobel Prize in 1975 for this discovery), all living organisms were thought to use only this process.

Reverse transcriptase enabled the understanding that certain life forms (retroviruses) have modified this process by allowing the flow of information from RNA to cDNA (complementary DNA). This is where the prefix for the name retroviruses originates—from backwards replication. Once the cDNA is made, the familiar flow of information from DNA → RNA → Protein resumes [☞ "Retroviruses;" "cDNA;" "Reverse transcriptase;" "Expression"].

Chargaff's rules

Erwin Chargaff (1905–2002) was born in Austria-Hungary (now Ukraine) and became a chemist during his studies in Vienna and at Yale. He worked in Berlin and Paris and ultimately became a Professor at Columbia University in New York City. In the mid 1940s, Chargaff turned his scientific attentions to the study of DNA, spurred on by the findings of Avery and his colleagues [☞ "Avery, Oswald T."] that DNA was a carrier (at that time thought to be the *only* carrier, long before the discovery of RNA-containing retroviruses) of genetic information.

Chargaff studied DNA from many different species and found that regardless of the source, the amount of adenine was virtually identical to the amount of thymine and the same was true for guanine and cytosine. Put simply, A = T and G = C. These became known as Chargaff's rules. While visiting Cambridge University in 1952, Chargaff discussed his findings with Watson and Crick. Although this conversation did not lead directly to the solution of the structure of DNA, it did provide information that was an important piece of the puzzle ultimately solved about a year later by Watson and Crick. To quote Crick, "...when you have one-to-one ratios, it means things go together." That A went with T and G went with C, per Chargaff's Rules, was central to the elucidation of the architecture of the double helix.

Chemiluminescence

Chemiluminescence refers to the emission of light from a chemical reaction. The phenomenon was first described in 1877.

One of the mainstay techniques of molecular pathology is labeling (through chemical attachment) DNA probes with chemical compounds that act as reporter molecules. When we hybridize chemiluminescently labeled DNA probes to DNA targets of interest and then perform the necessary chemistry, light is emitted—which reports to us a successful hybridization between target DNA (patient DNA in the clinical setting) and the probe we used. That emitted light can be "captured" on an X-ray film that we can study to help us answer questions about the nature of that patient's DNA. [☞ "Autoradiograph," keeping in mind that an "autorad" generated by a chemiluminescent probe is usually called a lumigraph or lumigram. Instruments called luminometers also exist; these "capture" emitted light and give us information about the kinds of analyses described here.]

Chemiluminescence occurs in nature, too, where it is called "bioluminescence." Examples include certain marine bacteria and the firefly.

Chimera (Chimerism)

Individual humans are composed of trillions of genetically identical cells. In other words, all the cells contain exactly the same DNA and were derived from the same zygote (the cell formed by fusion of the egg and sperm). People who are chimeras, however, are unusual in that they have two or more populations of cells that are genetically distinct. The name chimera is derived from Greek mythology (Greek "*chímaira*"), which describes the offspring of Typhon and Echidna as a female fire-breathing monster with the head of a lion, the body of a goat, and the tail of a snake.

An individual may become chimeric as a result of the fusion of two (or more) zygotes early in cell development (fraternal twin fusion). In this case, the resulting individual is a "mixture" of the two zygotes with some organs containing DNA from one zygote and other organs containing DNA from the other zygote. Other forms of chimerism can occur through blood transfusions, the sharing of blood by fraternal twins during development, fetal cell contamination of the mother, and experimentally, using laboratory recombination techniques (though not in humans).

Chromatin

Inside the cell's nucleus, chromosomes are tightly associated with specific cellular proteins. This protein-DNA complex is called chromatin. Even using a microscope, one cannot distinguish individual chromosomes within chromatin.

Chromosomal translocation [☞ "Gene rearrangement" and "Transcription"]

Chromosomal translocation is an abnormal occurrence. It is the exchange of portions of chromosomes, one with another, which is specifically referred to as reciprocal translocation. Another type of chromosomal translocation, called centric fusion, involves two complete chromosomes fusing to each other.

Some reciprocal translocations are known to be involved in the generation of certain cancers. The mechanism for this carcinogenesis is based on the movement of genes during the translocation from one "address" to another, causing the gene to escape its normal regulation by the cell, and leading to uncontrolled growth, namely, cancer. Chronic myelogenous leukemia (CML) is an example of a disease that results from a balanced chromosomal translocation, known as the Philadelphia chromosome (named for where it was discovered). The Philadelphia chromosome can be detected by cytogenetic [☞ "Cytogenetics"] laboratory analysis or more accurately and more often by molecular techniques like the Southern blot and reverse transcriptase polymerase chain reaction [☞ "Southern blot;" "cDNA;" and "PCR"].

CML is an example of a disease for which a specific drug has been developed based on our knowledge of the molecular pathogenesis of the disease. In the aberrant gene expression in CML described above, uncontrolled tyrosine kinase (TK) activity results. TK can stimulate cell growth. When it is not properly regulated, as in CML, the uncontrolled kinase activity is pathogenic for leukemia. A drug called Imatimib (trade name, Gleevec) inhibits the uncontrolled kinase activity, thus inhibiting the leukemia. Gleevec is one of a large number of new drugs in pharmaceutical company pipelines targeted against the molecular causes of disease. This is an example of why we are now in the exciting era of "genomic medicine." Herceptin for breast cancer, Iressa for lung cancer, and Avastin for colorectal cancer are similar examples.

Chromosome

Literally "colored body," referring to 19th-century scientists' microscopic observation of the ability of these bodies to be stained so strongly by blue and red dyes. In humans and all higher organisms, DNA is contained in tightly packed structures, called chromosomes, within the cell nucleus (see Table 2 for a comparison of chromosomes among species).

Chromosomes consist of DNA and proteins (virtually equal parts of histone and nonhistone proteins [☞ "Histone"]). The proteins help package the DNA by serving as a kind of scaffolding so that the very long DNA molecules can be condensed into a very small space. Humans have 23 pairs of chromosomes in every cell (except mature red blood cells, which do not contain DNA). See Figure 10.

Chromosomes are visually distinct only during cell division, a process called mitosis. When the cell is in that portion of the cell cycle where it is not dividing—the interphase period—chromosomes cannot be individually differentiated. Gametes, or sex cells (sperm and eggs), have half the normal complement of chromosomes (they are haploid) so that when they combine to form a fertilized egg, the full (diploid) complement of chromosomes (and DNA) that then goes on to form an embryo is present.

Table 2. Number of Chromosomes in One Cell of Different Species

Bacteria	1
Fruit flies	8
Peas	14
Bees	16
Corn	20
Frog	26
Fox	34
Cat	38
Mouse	40
Rat	42
Rabbit	44
Human	46
Chicken	78
Some species of fern plants	>1000

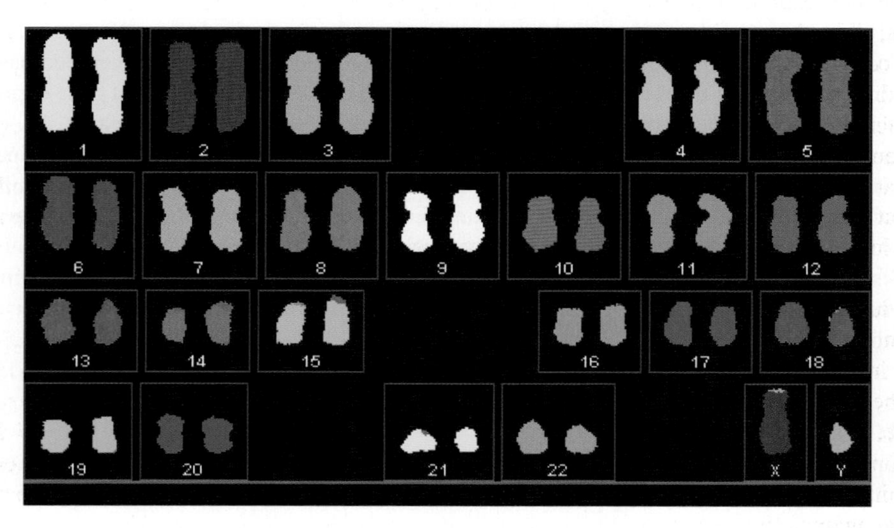

Figure 10. Colored karyotype. The 23 pairs of human chromosomes are pictured here. Note that there are two of each chromosome, including two sex chromosomes, x and y. The chromosomes are specially stained and photographed.

Image courtesy of Applied Spectral Imaging (ASI).

CLIA '88

CLIA stands for Clinical Laboratory Improvement Amendments of 1988. CLIA '88 is a U.S. federal law enacted to help ensure that all patient specimen results are accurate and reliable. Among other things, CLIA '88 defines the necessary qualifications for clinical laboratory directors and workers; requirements that must be met before a new clinical test is implemented in the laboratory; and aspects of laboratory quality control, quality assurance, and proficiency.

A clinical laboratory that performs testing on human specimens for diagnostic purposes must be CLIA-certified. Currently there are about 200,000 such certified labs.

They range from relatively small physician office labs, most of which perform only simple tests for which most CLIA requirements are waived, to large hospital-based and independent labs performing tens of thousands to millions of simple and complex tests annually, including molecular diagnostics and genetic tests.

Clinical decision point

The clinical decision point is a place within the assay range that indicates the need for a decision regarding patient management.

One area in molecular pathology where clinical decision points come into play is viral load assays. To help understand how a clinical decision point is derived, think of a blood alcohol level. For most individuals, a blood alcohol level at or below 0.07% is not considered unlawful, but in some individuals (novice drinkers, petite women, etc.), it may cause impairment. In other individuals, a blood alcohol level between 0.07% and 0.10% may or may not result in impairment, again depending on the individual. However, in all individuals, a blood alcohol level above 0.10% causes impairment in the *majority* of individuals and is considered illegal in all individuals. With this example, we can see that even though all individuals have a detectable blood alcohol level, their reactions are different, but there is a point, in this case 0.10%, where all individuals are considered impaired. This point is equivalent to the clinical decision point. Using the alcohol blood level as an example, we know that while a blood alcohol level >0.10% is considered legally impaired, experience tells us that there are some individuals who are not impaired, and that other individuals are impaired at lower alcohol levels, and that impairment is dependent on additional factors.

Similarly, there are clinical decision points in viral load testing. In HIV-infected individuals undergoing treatment, some do better (lower viral load) than others (higher viral load). Depending on the patient, a low viral load may cause problems, but in the majority of individuals, a low viral load indicates the patient is doing well. There is a point however, where the viral load is high enough to cause concern and possible intervention, for example, changing antiviral therapy. If the viral load is close to the pre-determined clinical decision point, the physician may or may not be concerned, depending on how the patient is doing. Or a threshold—the clinical decision point—may be crossed and a change in patient management implemented.

Clone; Cloned [☞ "Dolly"]

Three points:

1. **To clone a particular piece of DNA, including a gene, means to molecularly isolate it in the laboratory and insert it into a cloning vector.** A cloning vector is another piece of DNA that can be inserted into bacteria, viruses, or yeast cells, which then grow, simultaneously making many, many copies of the inserted vector containing the isolated piece of DNA. So to "clone" a piece of DNA means to isolate it and make more of it for study or use. An example of "use" is to molecularly clone the gene for insulin, insert it into an appropriate vector, and introduce this so-called recombinant DNA molecule (cloned DNA plus vector) into cells that can be grown either in the laboratory or on some large scale. During growth, the cells with the introduced insulin gene express that gene, thereby releasing insulin into the mix, which is purified and used medically.

2. **A clone of cells is a group of cells, all of which are genetically identical to each other.** In leukemia, for example, one white blood cell may escape the normal growth regulation the body imposes on its cells, due perhaps to a mutation caused by high-voltage electric fields, or overexposure to sunlight, or any one of a number of environmental insults. That cell becomes leukemic or cancerous and divides uncontrollably, thereby causing disease. All of the daughter cells that arise from that original leukemic cell are genetically identical and form a so-called monoclonal (one clone) population of cells.
3. **Science fiction becomes science fact, given enough time.** A favorite "scientific" cartoon depicts a little boy in front of his class in school during "Show and Tell." He is showing a frog to his classmates when the teacher reminds him that he was supposed to have brought something that he had made, to which the boy calmly replies, "I cloned her." Now that the cloned sheep Dolly [☞ "Dolly"] and many other cloned animals are reality, the cartoon is not so far-fetched.

CODIS (NDIS, SDIS, LDIS)

CODIS is the U.S. Federal Bureau of Investigation's Combined DNA Index System; NDIS is the National DNA Index System, while SDIS and LDIS refer to the State and Local DNA Index Systems, respectively.

CODIS combines computer technology with the power of DNA typing to discriminate among individuals (and thereby facilitate identification of perpetrators of violent crimes). CODIS stores and maintains DNA specimen data so that searches in support of forensic investigations can occur. Law enforcement professionals can search this information repository for, and/or provide, specific DNA information. Because of how CODIS is structured at the national, state, and local levels, different law enforcement agencies may cross-reference DNA information with that of other agencies in the U.S. In this way, DNA samples and their specific "genetic fingerprints" can be compared with each other to generate DNA matches and link what were previously unrelated cases.

By 1998, all 50 states had passed legislation requiring convicted offenders to provide samples for DNA databases. In October 1998, the FBI-maintained NDIS became operational. As of late 2007, there were almost 200,000 forensic profiles (DNA profiles from biological evidence found at crime scenes) and approximately five million convicted offender profiles in NDIS. Routinely, 13 different DNA markers called STRs (for short tandem repeats) are analyzed and used in CODIS; when combined to compare individuals, these markers provide enough power statistically to discriminate between any two people on the planet, except identical twins.

Short tandem repeats (STRs) of DNA are generally two to five base pairs in length. They exhibit length polymorphism among individuals. In other words, the number of times the STR CAT (for example) repeats itself may be three times in one individual and two times in a second individual: CATCATCAT vs. CATCAT. That length of polymorphism can help (in combination with more STRs) discriminate between two suspects or distinguish the suspect from the victim.

Codon [☞ "Anticodon;" "Genetic code"]

A three-base-pair sequence in DNA that encodes an amino acid; amino acids are the building blocks of proteins. When DNA is transcribed into RNA and then translated into protein by the cellular protein translation machinery, each amino acid in the growing protein chain is coded for by a specific sequence of three bases in the gene (DNA) that coded for that protein. Those three bases are termed a "codon."

Companion diagnostics

If you had tooth pain, you wouldn't want your dentist to implement the therapy, a root canal, until he or she had first done the "companion diagnostic" test, an X-ray of the tooth in question. This is a good analogy for the increasingly popular companion diagnostic type of testing occurring in molecular diagnostics laboratories.

Herceptin®, for example, is a therapy for breast cancer that is efficacious in patients with amplified levels of the HER-2/*neu* oncogene. Gleevec is an anti-leukemia medication (it has other applications too; [☞ table under "Pharmacogenomics"]) useful in patients with a specific chromosomal translocation. There are many more examples; tests should be conducted before the drug is prescribed and administered to learn if the patient is an appropriate candidate.

As pharmaceutical companies continue to develop drugs for diseases caused by particular molecular defects, it is increasingly important to test individuals for those specific defects to learn if they are indeed appropriate candidates for the drug; such information may have profound efficacy, side-effect, and economic ramifications. Indeed, it is likely that the FDA will mandate more and more companion diagnostic tests in conjunction with a possible prescription for the drug in question.

Complementary strands of DNA

DNA is an awfully polite molecule and the two sister strands are always complimenting one another. Actually, if you note the spelling of this entry, it refers to the nature of the DNA strands and not the canard that they're always praising each other.

The double-stranded DNA helix is made up of bases (among other things) and there are strict rules of complementarity that dictate how those strands pair up with each other. The rules are quite simple: the base adenine (A) always pairs, in DNA, with thymine (T); similarly guanine (G) always pairs with cytosine (C). So base pair complementarity, which stems directly from Chargaff's rules [☞ "Chargaff's rules"] dictates the sequence of the sister strand (or the daughter strand that is synthesized during DNA replication).

If one strand is:	5' - AGCTTTAAGTCGCTTA - 3'
then the complementary strand must be:	3' - TCGAAATTCAGCGAAT - 5'

You may also deduce from the above that the following statement about DNA is true: Within DNA the number of guanine bases = the number of cytosines; the number of adenines = the number of thymines; in other words, A = T and G = C. Erwin Chargaff discovered this truism in the 1940s.

In RNA, thymine is replaced by uracil (U). In RNA, a more or less single-stranded molecule, local regions of double-strandedness [☞ "Denature"] can occur and the base pairing is A to U and G to C.

Consensus sequences

A consensus sequence is the genetic equivalent of averaging numbers. Let's say that we are hired by a biotech company to determine the gene sequence that codes for a specific enzyme. We isolate the gene of interest from the DNA of a half dozen individuals and determine the sequence. When the sequences are compared ("aligning," molecularly speaking), we notice there are a few differences among the sequences. The project manager asks us for the sequence so we look at them and provide the *best* "average" of the sequences:

Sequence 1	ACTAGGCGT
Sequence 2	ACTAGGCGT
Sequence 3	ACTAGGCGA
Sequence 4	ACTAGGCGT
Sequence 5	ACAAGGCGC
Sequence 6	ACAAGGCGT
CONSENSUS	ACTAGGCGT

In the examples above, the underlined nucleotides in the third and last positions appear more often than the non-underlined nucleotides and are therefore the ones represented in what is referred to as the consensus sequence, something of an "average." A consensus sequence is also used when designing primers or probes to regions that are similar, but not identical.

Conserved

In molecular biology, the term conserved refers to bases or amino acids that do not change among all similar sequences that have been analyzed.

These may be single nucleotides that are conserved (position 210 in gene A is *always* a guanine); stretches of nucleotides (positions 5–15 in gene B *always* read ATTACAGTACG); single amino acids (a serine is conserved, i.e., is always present, in the sixth position of the protein [the codon may be different, because there are six codons that code equally well for serine]); or strings of amino acids [☞ "Codon"].

Conserved sequences may occur within a single species or across species. If a conserved sequence is found in many species, it is considered highly conserved. Conserved sequences are important because they indicate that the conserved nucleotide or amino acid is absolutely required for proper functioning of the gene product (if not, it would have changed during evolution because by definition it was proven to be non-essential).

Copy number variation (CNV)

Most genes on human chromosomes are present in two copies; one allele per chromosome. It has been found however, that large stretches of DNA, from thousands to millions of bases in length, vary in copy-number. These copy number variations (CNVs) may or may not include genes and if they do, then the "dosage" of a gene may be affected and may be present in unconventional numbers ($\neq 2$), which naturally can affect the amount of protein ultimately translated from those genes, possibly leading to disease.

As we were preparing this book, the following paper was published in the March 2008 issue of the *American Journal of Human Genetics:* "Phenotypically Concordant and Discordant Monozygotic Twins Display Different DNA Copy-Number-Variation Profiles." This could be a landmark publication. What does it mean? The authors of this study investigated the DNA of 19 pairs of monozygotic (identical) twins: twins that showed characteristics in common (phenotypically concordant) and not in common (discordant). The finding was that the twins, regardless of whether they were in the group with common characteristics or differing characteristics, showed copy number variations. Punchline? Identical twins are in fact NOT identical at the DNA sequence level. A lot of books and websites about genetics are going to have to be rewritten to take this new finding into consideration.

Cosmid

Cosmids are similar to plasmids in that they are extrachromosomal elements that are often used as cloning vectors. Cloning vectors are able to accept inserts of foreign DNA, infect a cell (in the laboratory), and replicate the DNA within the infected cell.

In this case, the cloning vector is derived from the bacteriophage lambda (λ), which infects the bacterium *E. coli* (bacteriophage: a virus that infects bacteria) [☞ "Bacteriophage" and "Lambda"]. Cosmids are so named because they contain the *cos* sequence from λ, which is required for "packaging" the DNA into viral particles.

Two main differences exist between plasmids and cosmids. The first is that cosmids are able to accept larger DNA fragments than plasmids. The second is that because they are derived from a virus, scientists can replace the non-vital viral genes with the DNA they are cloning and use the virus to infect the host cell. Once the DNA is inside the host cell it is replicated to large amounts for further analysis [☞ "Plasmid;" "BAC;" "HAC;" "YAC"].

Crick, Francis (1916–2004)

Francis Crick and James Watson (with a little help from their professional colleagues, principally Maurice Wilkins, Rosalind Franklin, and Erwin Chargaff) deduced the double helical nature of DNA and came to realize how that structure poised the molecule for its own replication. Crick and Watson shared the 1962 Nobel Prize for their work (along with Wilkins), went on to publish many more scientific manuscripts, write books, give international lectures on this and other subjects, become faculty, and head scientific research institutes. For their discovery, scientists affectionately call one strand of double-stranded DNA the "Watson strand" and the other the "Crick strand." Francis Crick's last position was that of J.W. Kieckhefer Distinguished Research Professor at the Salk Institute for Biological Studies in La Jolla, California, where he worked until his death in 2004. See Figure 11.

Cycle threshold (C_T) value

The cycle threshold (abbreviated C_T; aka crossing point or C_P) value is a term used in real-time PCR to designate when a positive reaction has occurred.

Real-time PCR uses fluorescent molecules to detect the formation of PCR product—in other words, more product equals more fluorescence. During real-time PCR, the fluorescent molecules emit a "background" signal. The threshold is set based on the level of

Figure 11. Watson and Crick (1953). James Watson (left) and Francis Crick (pointing with a slide rule) with their model of the DNA double helix. Note the hand-drawn figure on the wall.

background signal produced. When the fluorescent signal increases, because the target of interest is present in the patient specimen and is specifically amplified during PCR, over the background level, the cycle number where this occurs is defined as the cycle threshold (C_T) value. Readers of such a test result may now conclude that the specimen is positive for the presence of a particular virus or a particular mutation.

The C_T value can be used to determine how much DNA (or cDNA) was present in the starting sample. Since the fluorescent signal generated is proportional to the starting amount of DNA, the C_T value can be used to calculate the initial concentration. This is the basis for real-time quantitative PCR (q-PCR) [☞ "PCR;" "Real-time PCR"].

Cytogenetics

Cyto means cell. Because the genetics of the cell are dictated by the chromosomes, cytogenetics has come to refer to the study of chromosomes.

Specifically, microscopic techniques are used to visualize chromosomes. The microscopic visualization of the set of chromosomes is called the karyotype. Abnormal chromosome number and structure in particular disease states may be observed microscopically. Chromosomes can be translocated [☞ "Chromosomal translocation"], deleted, inverted, and increased or decreased in number; each of these may be diagnostic of a particular disorder. Two examples are (1) chromosomal translocations that occur in certain leukemias and (2) the increase in chromosomal number (an extra copy of chromosome 21) that results in Down syndrome.

A refinement of cytogenetics is called FISH (fluorescent in situ hybridization). In FISH, the DNA in chromosomes is denatured to cause hybridization to a fluorescently labeled probe. These probes, depending on the target to which they are complementary, hybridize to chromosomes in situ. In other words, hybridization occurs in the intact cell or in the spread-out area of chromosomes on a microscope slide. Areas of hybridization may then be viewed under a fluorescence microscope.

Changes in chromosomes, like amplification of certain regions sometimes seen in breast cancer, can thus be viewed. This is a common way in which the gene called HER-2/*neu* on chromosome 17 may be viewed to learn if it is amplified in cases of breast cancer. If so, the patient may be an excellent candidate for treatment with a drug, Herceptin®, targeted against the aberrantly unregulated action of amplified HER-2/*neu*. This is a specific example of pharmacogenomics: drug therapy specifically designed to target a particular genetic alteration in a particular disease [☞ "Pharmacogenomics"].

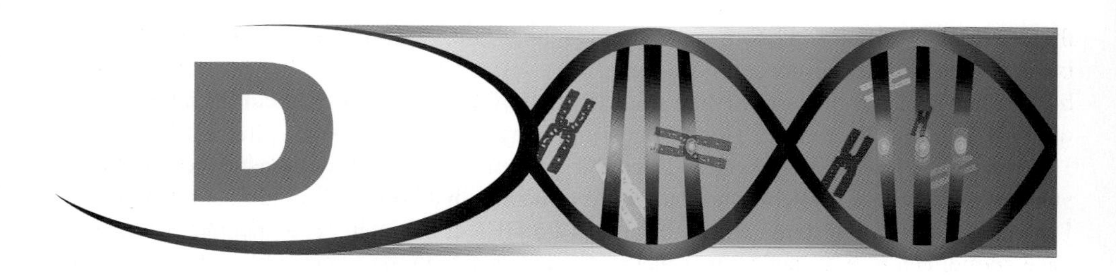

ddNTP

This is similar to the term used to denote dNTPs, except it denotes the dideoxyribonucleotides form of the nucleotide (missing the 3' hydroxyl group for continued chain elongation); ddNTPs are used in Sanger dideoxy sequencing methods [☞ "dNTP" and "DNA sequencing"].

Degenerate primers

We could go for the cheap joke here and say that degenerate primers are deviants that don't play well with others, but that would be too obvious. In actuality, degenerate primers are good at playing with others—other DNA sequences, that is. Normally, for a primer to be efficient at binding, an exact match between the primer and the template sequence is required. This exact match, however, is not always possible—not all genes are created equal. Degenerate primers compensate for these differences in sequence. They are a mixture of closely related, but not identical, sequences.

When would degenerate primers be useful? Let's say a physician orders a test to assess if a patient is infected with a particular virus. We know that this virus has several subtypes, but we do not know with which one the patient might be infected. Based on our knowledge of the viral genome we identify a region to amplify, but among the different viral subtypes we notice that the sequences are slightly different. For example, in subtype 1, the third nucleotide is an A (adenine), while in subtype 2, the third nucleotide is a C (cytosine). We design the primer to contain a mixture of sequences, some containing T (thymine, the complement of A) and some containing G (guanine, the complement of C). We then have a mixture of primers where 50% contain one sequence and 50% contain another sequence. These are known as degenerate primers, or mixtures of similar primers. Using this combination we have a better chance of amplifying the unknown viral subtype. In this example only one base was changed, but more than one base can also be substituted.

Degenerate primers may also be the method of choice when the sequence of interest is not fully known or when designing primers based on amino acid sequence, where the amino acid code is used to derive possible primer sequences. Because the genetic code is redundant (more than one codon can code for the same amino acid) degenerate primers would be needed if using this approach to amplify a target using only the known amino acid sequence.

Denature

DNA is naturally double stranded. Often in the (clinical or research) laboratory, in order to work with DNA or detect a specific feature of it (for example, a mutation), we must make the DNA single stranded so that we can get at the sequence of interest. To "denature" DNA is to make it single stranded. This is most often done by heating the DNA solution to near-boiling temperatures (near 100 °C), or by treating the DNA with strong alkali (the opposite of acid).

RNA is a single-stranded molecule, but it may exhibit local regions of double-strandedness. For example, see the picture of tRNA (Figure 2) under "Anticodon." Double-strandedness in a mostly single-stranded molecule can occur due to particular base sequences in a given RNA molecule. Sequences may be such that under the right biochemical conditions, portions of the strand come into close enough proximity to other portions of the strand to pair up. For instance, if these two regions exhibit sequences with complementarity [☞ "Complementary strands of DNA"], base pairing occurs. This configuration, where parts of the RNA molecule are base-paired with other parts of the molecule is known as an RNA secondary structure [☞ "RNA"].

It is important to remember that RNA may be depicted as a straight line on paper for ease of illustration, but in the cell there is a three-dimensional aspect to all molecules. This three-dimensional aspect may make neighbors of regions of nucleotides with complementarity.

DNA

Deoxyribonucleic acid [☞ "Nucleic acids;" "Nucleotide;" "RNA;" ☞, in fact, every entry in this book]

"The stuff of life," DNA is the genetic material that is passed from parent to progeny and propagates the characteristics, in the form of the genes it contains and the proteins for which it codes, of the species. That DNA is the carrier of genetic material was shown by classic experiments of Griffith and then Avery, MacLeod, and McCarty [☞ "Avery, Oswald T."]. Until that point—about 60–70 years ago—conventional wisdom suggested that proteins were the only macromolecules complex enough to carry genetic information (Figure 12).

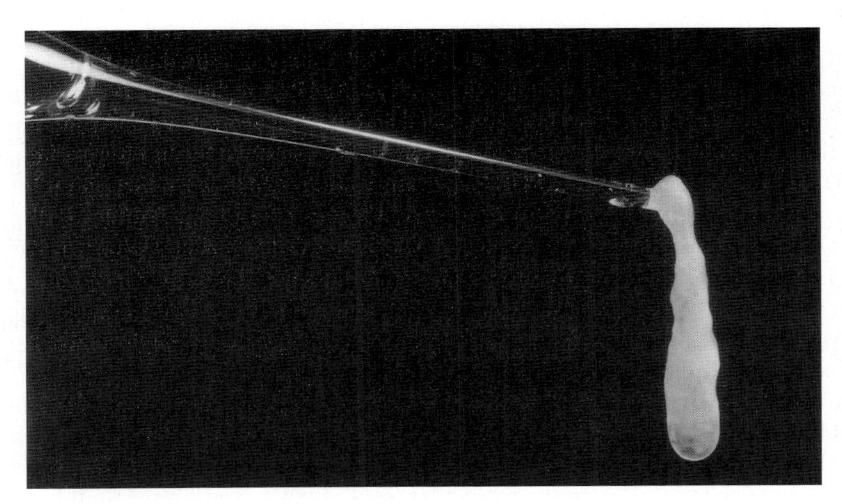

Figure 12. Purified human DNA hanging from the end of a glass rod (the liquid you see is alcohol).

DNA bank

This is a service just like a sperm bank, tissue bank, or traditional financial institution where money is stored. Many institutions house DNA banks where DNA extracted from certain patients' tissues (at the patient's request with medical advice, of course) is frozen and stored indefinitely. In this way, the DNA is available if, for some reason, it needs to be tested in the future.

Why might such testing be necessary? The examples are not the most pleasant circumstances to consider. For example, identification of remains may become necessary, and matching DNA profiles obtained from remains with banked DNA on a known individual may provide a basis for absolute identification. In fact, this has been exploited by the military, and the first Gulf War was the first in our history where no interments were made in the Tomb of the Unknowns in Arlington National Cemetery.

From a medical point of view, DNA banking could provide definitive diagnoses of some genetic disorders that may be difficult or impossible to diagnose. For example, children may present to clinics with rare, unrecognized, or unique ailments. A clinical diagnosis may be suspected, but there may be insufficient laboratory data upon which to base a definitive diagnosis. No confirmatory lab test may be available for the condition at the time. In the future, however, such patients may be provided with a definitive diagnosis through research that has led to DNA diagnostic tests. The banking of DNA from these affected patients, who may not survive, may permit definitive diagnosis and recurrence risk counseling for the parents of these patients in the future. The availability of such DNA may also be of value in the counseling of siblings and other family members.

The value of DNA banking is not limited to such unfortunate circumstances. A stored DNA sample of a parent or grandparent may be of value to descendants of that individual with respect to counseling concerning reproductive and health issues.

DNA chips

We don't believe we'll be seeing this as the latest offering from Frito-Lay anytime soon. In fact, with the public concerns over transgenic [☞ "Transgene" and "Transgenic"] corn found in Taco Bell® taco shells in 2000, DNA technology is something from which such companies would most likely distance themselves. As DNA technology is exploited for the betterment of humankind, lines between social and biological sciences will sometimes blur. Questions that arise are best discussed and ultimately answered by an informed populace.

DNA chips are an attempt at DNA miniaturization, which is ironic (isn't it?) because DNA is already small. DNA chips are ordered arrays of oligonucleotides ("oligos" for short; ☞ "Oligonucleotides"). Illumina and Affymetrix (affectionately known as Affy in the trade) are the two companies that first come to mind when the term "DNA chip" is used. Other companies compete in this space as well, including one called Luminex, though its "chips" are more appropriately thought of as arrays since they are liquid-phase based, not solid-support based like Affy's or Illumina's chips.

Chips and their "cousins," liquid-based arrays, are starting to catch on diagnostically. Complex diseases like cancer have gene expression "signatures" demonstrated by chip-based analysis that may be exploited diagnostically and for therapy choice, by targeting disease-specific anomalies on the molecular level. Arrays exist for respiratory panels of pathogens, to pinpoint what might be the causative agent so that the most appropriate therapy can be rationally applied. Drug metabolism can now be assessed by investigating genes and their variants responsible for appropriate or inappropriate metabolism of drugs; in this way, adverse reactions and non-efficacious dosing can potentially be avoided.

For example, a specific gene (cyp450 *2D6*) may be interrogated before using tamoxifen in breast cancer patients. With this information physicians can learn which patients may be poor metabolizers of tamoxifen, and which patients would more likely benefit from a tamoxifen alternative. Furthermore, some drugs inhibit the very enzymes the body uses to metabolize tamoxifen to the active metabolite that fights breast cancer. It is valuable to learn a patient's cyp450 genotype in order to more rationally manage breast cancer and its sequelae using various drugs. Learn more at http://www.mayomedicallaboratories.com/media/articles/communique/mc2831-0107.pdf.

In the genomics space (in other words, high density arrays where one is using chips for discovery work; such discoveries can lead to markers against which diagnostic tests and/or drugs can be developed), chips exist that allow investigation of up to approximately a million single nucleotide polymorphisms [☞ "SNP"]. Because there are three billion base pairs in the genome, this represents one polymorphism (potentially informative difference between individuals) about every 3,000 bases. Chips also exist to allow investigation of about 50,000 transcripts. How can there be about 23,000 genes and about twice as many transcripts? Genes are differentially transcribed. In other words, one gene can lead to multiple RNA transcripts and thus multiple gene products (proteins).

Think of DNA chips and arrays as the way to do many biological reactions simultaneously, just as computer chips do many mathematical calculations simultaneously (see Figure 13).

DNA computers

In principle, DNA can be used for information storage or even computation. In fact, DNA could be the basis of the smallest computers yet invented. The base pair coding of G-A-T-C means that each base pair in the helix could represent, for example, a zero (G or A) or a one (T or C). In this way DNA molecules could be used to store information. This is very similar to how nature uses DNA to encode amino acids for protein synthesis. Scientists have also proposed using DNA as a computation engine, like the processor in your computer, but one that is much, much smaller. A DNA computer would exploit the molecular nature and biochemistry of DNA to manipulate sequences of base pairs in defined ways to achieve computation. While DNA information storage or DNA computation is possible in theory, there are many hurdles to overcome. It will be many years before we know if DNA computation is practical. (Thanks to Dr. R. Levine for providing this entry.)

DNA fingerprinting

[☞ Paternity testing]

DNA labeling

Through a variety of biochemical manipulations that rely on the action of proteins or chemical modification, DNA molecules can be "labeled." "Labeling" means tagging a DNA molecule with a small reporter molecule that can be detected using an X-ray film (if the

Figure 13. Starting from lower left and moving clockwise around the perimeter: The large green plastic cassette contains an array from GeneMaster, Inc., designed to analyze specimens for causes of meningitis at the DNA level. Above it is a NanoChip® electronic microarray from Nanogen, Inc., which contains a tiny silicon chip on which 100 test sites may be used for different molecular diagnostic assays. Next is an Intel Pentium computer chip. To its right is an Affymertix high density GeneChip® array that may be used to assess over 47,000 human transcripts. To its right is an "A-ring" from Roche Molecular Diagnostics that is used in a specialized instrument to amplify up to 12 different patient specimens for a specific target; for example, the genome of Human Immunodeficiency Virus. Beneath the "A-ring" are three iterations of a medium-density (36 targets) array from Clinical Micro Sensors, a Motorola company, used in bioelectronic detection of DNA targets such as cystic fibrosis-causing mutations. Next is the DNA LabChip® from Agilent Technologies that is used to automate laboratory manipulation and analysis of DNA. Next to this is a high-density array that is part of the CodeLink platform, originally manufactured by Motorola, ultimately sold to GE and since discontinued. Beneath the coin is an ultra-dense, low-power, lower-cost memory chip being developed by ZettaCore, Inc., for the microelectronics industry. Above the coin are three more items: the black rectangular silicon chip is manufactured by Sequenom, Inc., and incorporates a high-density, photo-resistant array of mass spectrometry DNA analysis sites; next to it is an unusually shaped (this shape promotes rapid analysis) laboratory tube from Cepheid, Inc., used in its real-time amplification applications for clinical diagnostic purposes; for example, detection of Streptococcus infections in pregnancy; last, is shown a glass capillary tube from Roche Applied Sciences used in real-time amplification of specific nucleic acid targets in clinical or research specimens—the instrument that accommodates these capillaries can process up to 32 at a time.

reporter is radioactive or luminescent) or, more commonly, additional chemical reactions that generate visible color or a fluorescent signal, generally the "end game" in a clinical laboratory test exploiting DNA or RNA.

The image we see on the film or what we view by means of observing color or a fluorescent signal allows laboratorians to answer questions about a particular sample of DNA. Examples of these questions include the following:

- Is a particular microorganism's DNA present in this patient sample?
- Is a particular mutation present in this DNA sample?
- Is there a particular genetic marker present that indicates the presence of cancer, or indicates paternity?

One of the most practical applications of DNA labeling is in the process known as "real-time PCR," described later in this book. In real-time PCR, the use of fluorescently labeled DNA molecules allows detection of the amplicons as they are generated during amplification [☞ "Autoradiograph;" "Chemiluminescence;" "PCR 'wannabes';" "Real-time PCR;" "Southern blot"].

DNA probe

[☞ "Probe"]

DNase

[☞ "Nuclease"]

DNA sequencing (and HIV-1 genotyping)

[☞ "Agarose;" "Electrophoresis;" "Human Genome Project"]

In the mid-to-late 1970s and on into the early '80s, techniques were invented and refined to read the sequence present in a particular piece of DNA. Throughout the 1990s the process was scaled up to the point where we were able to realistically contemplate sequencing all three billion bases in the human genome. This was in fact completed in 2003 [☞ "1995"].

When an earlier version of this book was written, the sequencing of the first dog genome was underway; it was finished in July of 2004 [☞ "Boxer"]. Since then, full sequencing of genomes from diverse species has flourished, such that we could literally write a book called *DNA Genome Sequences from A to Z*. Full-genome sequencing from *Aspergillus* (a fungus among us) to the zebrafish are completed or underway.

DNA sequencing has become so automated and robust that whole genomes can now be sequenced in days, even hours in the cases of small genomes. The day will likely come when one's own genomes can be sequenced for around $1,000 and the information carried on a kind of credit card, CD, or on your iPhone or BlackBerry for medical analysis at appropriate times; think of this as the future of Medical ID-alert tags, alerting emergency response teams to diabetes or penicillin allergy. Future emergency personnel may "wand" a chip or program that everyone carries to alert them to all sorts of vital medical information that would be useful to know in an emergency.

Of course, there are medical, ethical, business, and societal issues attached to such new technology. It is not a question of *whether* we will have to address those issues but *when*; this is indeed so because technology marches on more quickly than society can adapt.

TECHIE: In the early day of DNA sequencing, the technique depended on DNA electrophoresis in high resolution polyacrylamide gels (not unlike agarose gels in principle), which were also called sequencing gels. Sequencing or slab gels are capable of resolving single-stranded oligonucleotides, hundreds of base pairs in length, that differ in size from each other by just a single deoxyribonucleotide. Enzymatic or chemical reactions make oligonucleotide products (encompassing the region of interest) that end with either adenine (A), guanine (G), thymine (T), or cytosine (C), the four deoxyribonucleotides that make up DNA. The oligonucleotide products of the reactions are then electrophoresed in adjacent lanes of a sequencing gel. The one base pair resolution capability of these gels allows one to "read" from the gel the sequence of the DNA being interrogated.

TECHIE: There are two traditional methods for performing DNA sequencing: (1) dideoxy (Sanger) sequencing and (2) chemical (Maxam-Gilbert) sequencing. In dideoxy sequencing, unusual bases called 2', 3' dideoxyribonucleotides (ddNTPs) are used as substrates for growing oligonucleotides chains synthesized using the DNA of interest as a template. When a ddNTP is incorporated, oligonucleotide chain growth is blocked, because that chain now lacks a 3' hydroxyl group for continued chain elongation. Four separate reactions are run, each with a unique ddNTP. Manipulation of the ddNTPs:dNTPs ratio results in chain termination at each base occurrence in the DNA template corresponding to the included ddNTP. In this way, populations of extended chains exist within each reaction with differing 3' ends specifying a given ddNTP—in other words, specifying the sequence.

Present-day DNA sequencing and DNA fragment analysis have been refined from conventional slab gel electrophoresis to platforms that depend on capillary electrophoresis (CE) technology. CE systems provide more consistent and standardized results more quickly and with less effort than slab gel electrophoresis. Commercially available CE instrumentation has automated nearly every step in sequencing: DNA denaturation, sample injection, electrophoretic separation, data analysis, and database interaction steps. Sequencing via slab gel electrophoresis takes many hours to perform, whereas CE can generally be done in less than an hour while using less sample.

Electrophoresis is the application of an electric field to molecules like purified DNA or proteins to cause their migration and physical separation based on the molecules' inherent physical characteristics (like charge and size). In CE, electrophoretic separation occurs in very thin capillary tubes several inches long but only 50–75 microns in diameter. DNA fragments separate based on differences in mass. Because DNA has a net negative electrical charge, it migrates towards the positive node (the anode) of the capillary when an electric field is applied to an electrolyte solution within the capillary. The DNA migrates within the capillary through a polyacrylamide gel matrix that acts as both the electrolyte solution and a molecular sieve. Excellent automated and semi-automated DNA sequencing equipment is available from companies like Applied Biosystems, Siemens, and Beckman Coulter.

TECHIE: Newer technologies such as pyrosequencing and polony sequencing have sped up the process even more. Pyrosequencing is a method based on release of the pyrophosphate byproduct, which is generated during nucleotide addition to the growing chain. Recall that the added nucleotide is a triphosphate (dNTP; where T = triphosphate). During DNA synthesis, two phosphate molecules are released and only a monophosphate is added to the growing DNA strand. It is the two released phosphates, in the form of pyrophosphate, that ultimately generates a luminescent signal. The identification among the four bases is based on adding them sequentially, so that if dATP is added and a signal is produced, then we know that an A was added to the growing DNA strand. (If two or more adenines are added, the signal strength is doubled or tripled.) Pyrosequencing is fast; approximately 100 million nucleotides can be analyzed in a single eight-hour shift. The length of the synthesized product, however, is much shorter (~200–400 bases) than can be achieved with dideoxy sequencing, meaning that many more reactions are required to fully sequence the final product.

Polony (POLymerase colONY; discrete clonal amplifications of a particular DNA molecule to be sequenced) sequencing is another unique and fast-sequencing method that involves immobilizing DNA fragments, amplifying them in place, and then sequencing the amplified "clones." Sequencing is performed by adding the four differently labeled nucleotides to the reaction, washing away unincorporated nucleotides, and checking to see which nucleotide was retained. This cycle of nucleotide addition, removal of unincorporated nucleotides, and signal capture is then repeated throughout the chain synthesis. The method is technically easy and provides ultra-high throughput.

From the molecular point of view, DNA sequencing is the "gold standard" for detection of mutations and relevant DNA sequences. The most common application of DNA sequencing in the clinical laboratory early in the 21st century is HIV-1 genotyping. Mutations in the HIV-1 genome occur under the influence of anti-retroviral therapy. Patients failing therapy (as demonstrated by another popular molecular diagnostic test that quantifies the viral burden of HIV in a patient's bloodstream) may undergo HIV genotyping to learn what mutations are present in the infecting viruses. Some of these mutations render HIV resistant to certain drug therapies. Knowing the HIV mutation pattern in each patient allows infectious disease physicians to optimize drug selection to treat that patient in the most rational manner—in other words, with drugs to which the virus is sensitive. Here's what's implicit in this description: that we understand enough about the HIV genome and its biology to know which mutations render the virus resistant or sensitive to the drugs we use to combat HIV infection.

DNA sequencing has been relevant, both directly and indirectly, in virtually every disease diagnosable at the molecular level. We use the word "indirectly" because even though a particular disorder may be detectable by molecular methods other than sequencing, the success of a particular technique, e.g., PCR, depends on knowing the sequence of the gene of interest so that appropriate primers for PCR can be synthesized [☞ "PCR" and "Primer"]. See also Figure 14.

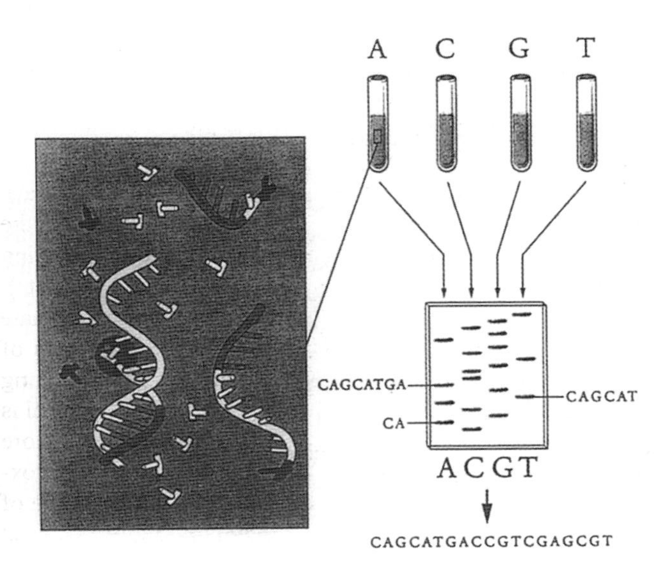

Figure 14. Ingredients for sequencing reactions are similar to those for PCR—template DNA, primer, DNA polymerase, and all four nucleotides—plus a small proportion of an additional chain-terminating nucleotide (A in this example). The reactions create a population of fragment sizes, all ending in a given letter. Read the sequence from the lightest to the heaviest band of DNA in the gel photo.

DNA/RNA extraction and purification

This is the process of extracting (releasing) and purifying either DNA or RNA from tissue, cells, or viruses.

Tissue examples could include whole blood, bone marrow, cerebrospinal fluid, or solid tissues such as lung, liver or tumor. Every cell in the body is nucleated (except mature red blood cells) and contains DNA, as well as several types of RNA. Every tissue therefore may be considered a suitable candidate for DNA or RNA extraction. Blood is obviously an excellent source for DNA or RNA due to its accessibility and the fact that it is full of white blood cells and other DNA/RNA-containing cells. Bacteria and fungi also contain DNA and RNA, while viruses contain either DNA or RNA, not both.

DNA and RNA extractions, using different methods and chemicals, are routinely performed in the clinical molecular diagnostics laboratory. DNA is a fairly stable molecule; RNA, on the other hand, is much less stable or more "labile" than DNA and special procedures must be undertaken in the laboratory to ensure that it is recovered intact from starting material.

Keep in mind that the right specimen is a key consideration in obtaining the nucleic acid to be used in a diagnostic test. One would not extract DNA from blood to look for bacterial DNA if one suspected a urinary tract infection; in this case, urine would be the correct specimen of choice. Similarly, diagnosis of recurrent leukemia would be something examined in blood or bone marrow and generally not in another tissue. If therapy proved successful and very few or no leukemic cells are present, sampling error could be a problem. (Sampling error is a way of expressing the idea that there may be, for example, ten or a hundred leukemic cells in a recovering patient's entire bloodstream, which is, of course, made up of liters and liters of blood.) If one samples a small volume of the patient's blood, for example, ten milliliters, it is statistically possible that the sample taken may simply contain no leukemic cells and a false negative laboratory test result would be obtained.

TECHIE: While it is not the intent of this book to overwhelm you with laboratory protocols, all DNA/RNA extraction and purification protocols have several basic steps in common. The first step is to release the nucleic acid (DNA or RNA) from the cell or viral particle—this is the "extraction" step.

Several methods can be used to accomplish this step, and the method of choice is based on several factors, such as starting cell type. Chemicals, heat, sound waves (sonication), or just really rough handling are commonly used. The goal of this step is to break open the cell so the nucleic acid can be "purified" away from the rest of the cellular components (membranes, proteins, and other small molecules).

Purification is usually accomplished by separating the nucleic acid and "washing" away the unwanted components. Centrifugation is one way to separate the nucleic acid. In this case, the nucleic acid is collected in the bottom of a test tube and the supernatant containing the unwanted components is removed. Another way is through binding of the nucleic acid to something (silica or magnetic particles coated with oligos work well) followed by washing [☞ "Oligonucleotide"].

The last step is to release the purified nucleic acid in a process called elution. This simply means releasing it from either the bottom of the test tube (in the case of centrifugation), or off the silica or magnetic particles.

Many fine commercially available kits may be used for nucleic acid purification. Furthermore, many automated, robotic solutions also exist for DNA and RNA purification from various specimens.

dNTP

This term denotes the four deoxyribonucleotides: dATP, dCTP, dGTP and dTTP. It is used generically when referring to any one of the four nucleotides, not to a specific one in particular. The N in dNTP comes from the IUPAC code for aNy nucleotide. The term dNTP is commonly used when designating that all the nucleotides are present in a reaction, such as in a PCR or DNA sequencing reaction [☞ "DNA sequencing;" "PCR;" "IUPAC"].

Dolly (1996–2003)

"Dolly" is the name of a unique lamb born in 1996 who died in 2003 of a lung infection common in older sheep, especially those housed indoors (there is some evidence to suggest that her somewhat premature death was a function of the fact that she was cloned). She was not a product of normal sexual reproduction between a ram and an ewe. Rather, Dolly was "created" by genetic manipulation. An unfertilized sheep egg cell was the starting material—the vessel, if you will allow that analogy. The naturally present nucleus within the egg cell was microsurgically removed. In its place was substituted the nucleus from an adult sheep mammary (breast) cell—the vessel's new contents. That laboratory-manipulated cell, after brief growth in a Petri dish, was introduced to a surrogate mother's uterus for implantation and "normal" pregnancy. The resultant birth yielded a cloned sheep, Dolly. The huge amount of media coverage cemented the term "cloning" in the minds of the public, and admittedly, "cloning" has a little more cachet than the more scientifically descriptive term, "somatic cell nuclear transfer." Note that more than 250 attempts at cloning were necessary before a sheep (Dolly) was successfully cloned.

A mammary gland (breast) cell was used for the donor nucleus. The transferred nucleus contained all the genetic material that would have been present in a fertilized egg. The process removed the father from the equation. Dolly is thus an exact duplicate, at the molecular level, of the adult sheep that contributed its DNA to form Dolly. Dolly was cloned from the donor sheep.

Since Dolly's birth, there have been other cloning attempts, some successful, some not. Prometea was the first cloned horse (it took more than 800 attempts), born in Italy in spring 2003. She was formed by fusing an adult skin cell and an empty equine egg cell donated by an adult female. The fused cell was then reintroduced into the same female who went on to deliver a healthy, genetically identical foal. It makes one wonder if owners of famous horses like Secretariat, Man of War, and Seabiscuit have stored blood or other cellular samples of these great champions for their subsequent cloning. In fact, a French company, Cryozootech, has been established to take advantage of the possibility of cloning championship horses.

ABS Global (De Forest, Wisconsin) has successfully cloned cows and pigs. ABS (whose cloning is more aptly described by the term "cell fusion" rather than nuclear transfer) is responsible for a bull named "Gene" (clever, huh?) and uses genetics in animal husbandry. The application of cloning to the human transplantation field is an exciting possibility. Consider cows with neurological tissue compatible (no tissue rejection) with humans to treat Parkinson's disease, for example. (By the way, transplantation across genus and species boundaries is called "xenotransplantation.")

The ethical ramifications of cloning are considerable. For example, could the pain of losing a loved one be mitigated by cloning that person before his/her death? Of course, a clone would not possess the life experiences of the original person. But what if we're talking about a terminally ill newborn? These are important concerns to be taken seriously. Human therapeutic cloning, or the technology to clone organs needed for transplantation, is being refined. Such technology could theoretically lead to the ethically difficult notion of a cloned person. It is the same technology however, that could have such valuable consequences in therapy for human disease through generation of appropriate healthy cells that could be used to replace diseased cells. That further research into animal and human (therapeutic) cloning will very likely yield medical benefits must be considered in the necessary public debates about cloning.

As of early 2008, the following had been cloned:

Carp
Cat
Cattle
Deer
Dog
Ferret
Fruit flies
Gaur
Goat
Horse
Mice
Mouflon
Mule
Pig
Rabbit
Rat
Rhesus monkey
Sheep
Water buffalo
Wolf

Dominant

[☞ "Inheritance"]

Downstream [☞ "Upstream"]

Just downstream from the "Dolly" entry in this book, you'll find the entry for "downstream." It means towards the 3' (right-hand) end of the DNA molecule, or "just down the road a piece" from a particular point. So if a nonsense mutation in a gene that will prematurely STOP protein synthesis is introduced just downstream of the gene's transcription initiation site (TIS), then that mutation is just a few bases away:

TIS--------------------MUTATION--------------------NORMAL END

In this example, the MUTATION is just downstream from the Transcription Initiation Site (TIS). If the MUTATION were closer to the NORMAL END of the gene, it would be further downstream of the TIS. If the mutation is very close to the normal end, enough of the correct protein structure may be retained such that the mutation is not particularly deleterious or harmful. In this latter example, the TIS is just upstream from the mutation.

Dropout allele

Yes, even in the world of molecular biology there are dropouts. In this case, however, a dropout is not a fourth-year resident who decides to become a professional NASCAR driver (yes, that actually happened), but rather is the failure of one of the two alleles present in a heterozygous cell to amplify. Why is this important?

Recall that all cells carry two sets of chromosomes, one set (23 individual chromosomes) from the mother and one set (23 individual chromosomes) from the father. Also recall that the chromosomes contain the same alleles (gene forms), but that these alleles may not be identical. Now let's say that one parent has a mutation that increases his/her risk for developing coronary artery disease and we would like to know if this mutation (on an allele) has been inherited by the offspring. We perform a test to analyze this, but one of the alleles fails to give a detectable result. This is known as a dropout and presents a technical problem for performing and analyzing the test results. Careful development of newer laboratory tests has helped to eliminate this problem.

Duplex

Depending on where you learn your real estate, duplex refers to two homes or domiciles that are one on top of the other or side by side. This is similar in DNA chemistry, where a duplex refers to two nucleic acid molecules that are bound together such as seen in good old double-stranded DNA. Duplexes can be DNA:DNA; RNA:RNA; or DNA:RNA hybrids.

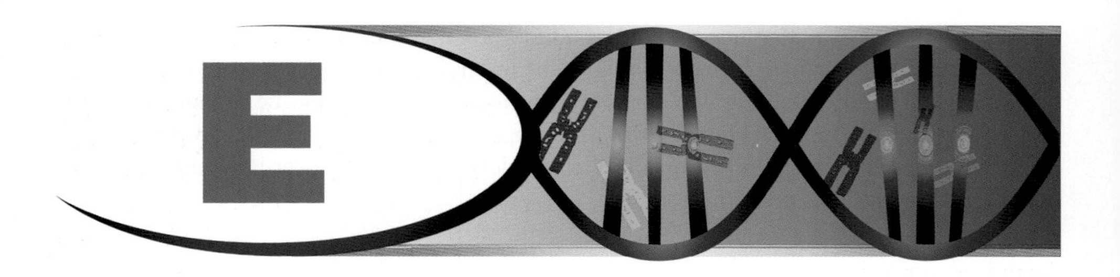

Electrophoresis [☞ "DNA sequencing"]

Electrophoresis is a commonly used technique, in both clinical and research laboratories, and exploits the migration of DNA and proteins in an electric field (Figure 15).

DNA migrates in an electric field inversely proportional to its molecular weight. That's a fancy way of saying the heavier (or larger) the piece of DNA, the more slowly it migrates while the lighter (or smaller or less massive) the piece of DNA, the more quickly it migrates (try running fast carrying a 50-lb. bag of dog food).

Typical agarose DNA electrophoresis in the laboratory proceeds like this: A molten gel is poured [☞ "Agarose"] and cools into a semi-solid material. It is next overlaid with water, to which has been added the right salts and chemicals (called electrophoresis buffer), and a DNA solution (the DNA may or may not have first been cut into smaller fragments; ☞ "Restriction Endonucleases"). Mixed into the DNA solution are a dye (generally blue) and something to increase the density (a sugar like sucrose or perhaps glycerol). The dye accomplishes two things: (1) the DNA solution is made blue so it can be made visible when applied to the gel, and (2) the blue dye allows the progress of electrophoresis to be monitored, because it also migrates in the electric field. The sucrose or glycerol makes the DNA solution denser than the water-based buffer, allowing the

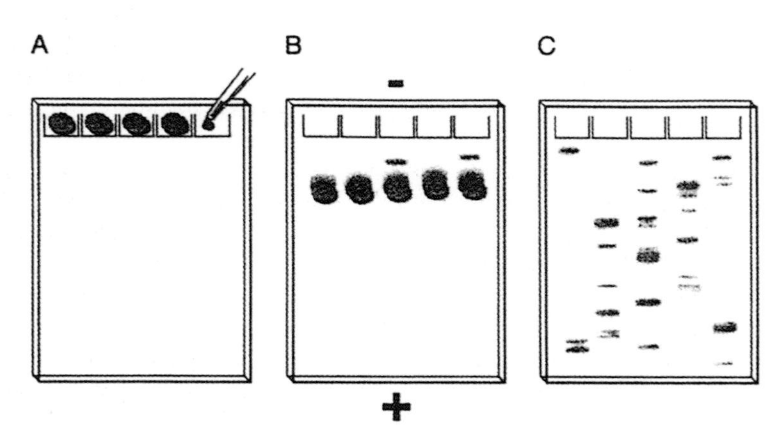

Figure 15. Electrophoresis. (A) Sample solutions, which are mixtures of molecules of various sizes, are "loaded" into wells in a special gel matrix. (B) DNA has a net negative charge, so when electrical current is applied to the gel, the DNA fragments migrate into and through the gel toward the positive electrode. (C) Various methods can be used to visualize the bands that are generated at the completion of electrophoresis. Smaller molecules migrate more quickly through the gel and form the bands near the bottom. Larger molecules move more slowly and remain closer to the top.

DNA solution to stay inside the well of the gel; absent sucrose or glycerol, the water-based DNA solution would float away when the water-based buffer is added.

Now that the gel has been successfully loaded, electrical leads are attached to each side of the gel box and to a power supply, which generates an electrical current of 20–250 volts. At this point you can usually go home or do something else in the laboratory. Electrophoresis can proceed from as little as 10 or 15 minutes to overnight, depending on what question is being investigated. The reason this works is that an electrical circuit between the gel and the power supply has been created, and as electricity flows through the buffer used to overlay the gel, it carries along the DNA molecules which were loaded (and the blue dye).

The progress of electrophoresis can be monitored by watching the progression of the blue dye. This is only a little bit more exciting than watching paint dry. The blue dye also serves another purpose and that's to make sure the polarity was not reversed when the electrical leads were connected. If that error was made, both the DNA and the dye will go in the wrong direction and the work will be ruined. Coming back a couple of minutes after starting the electrophoresis ensures we aren't practicing the dreaded laboratory error of "retrophoresis." If this has occurred, the power can be turned off, the leads changed to the correct positions, and the electrophoresis will proceed properly.

There are various uses for DNA electrophoresis, e.g., DNA sequencing, DNA fingerprinting, DNA quality assessment, and DNA restriction fragment analysis.

ENCODE

ENCyclopedia of DNA Elements. The human genome has been completely sequenced. Often, the analogy we use is that we have spelled out the letters in the "book of life." We know that there are 24 chapters (22 autosomal chromosomes plus the X and Y chromosomes) but we haven't yet figured out all the punctuation within each chapter.

Allthewordsruntogetherwithoutanyspacingandsoitshardtounderstandwhattheymean. We don't know which words are really important (because they're genes, for example) or which are unimportant (like junk or intronic DNA, for example). Indeed, we are finding that even so-called "junk DNA" may have a regulatory purpose within the genome [☞ "Junk DNA"]. We don't know which stretches of "words" suggest a connection to another stretch elsewhere in the "book."

The National Human Genome Research Institute (www.genome.gov) is leading a project to identify all the important and functional sequences in the human genome. Examples of "important" sequences include those that code for genes, promoters, transcriptional regulatory elements (think of these as TV or iPod volume controls), and determinants of chromosome physical structure. The research funded under this initiative will provide us with an encyclopedia of DNA elements, beyond "mere" genes. This supplemental information will help us exploit the basic knowledge inherent in having completed the human genome sequence and allow us to use that knowledge to better understand the genome, disease, disease risk, and diagnostic and therapeutic targets. Ultimately ENCODE will enhance the annotation to the human genome and lead to better understanding of our own biology.

Enhancer [☞ "Expression"]

Enhancers are stretches of bases within DNA, about 50–150 base pairs in length, that increase the rate of gene expression.

Enhancers have particular sequences of bases that are recognized and bound by different DNA binding proteins. These proteins act in different ways to regulate the expression of genes, for example "up-regulation" and "down-regulation" where genes are transcribed into mRNA in greater and lesser abundance, respectively. Enhancers may be physically close to or far from the gene they are responsible for regulating. When enhancers are thousands of base pairs away from a gene they regulate, there are experimental data to demonstrate that the three-dimensional structure (supercoiling) of DNA may bring enhancers and genes in physically close proximity.

Viruses often carry enhancer sequences in order to get their genes expressed preferentially and in higher concentrations over normal cellular genes. Mutations in some tumor cells also act as enhancer elements and cause over-expression of genes involved in cellular proliferation, growth, and differentiation.

Enzyme [☞ "ase"]

Enzymes are the biochemical tools of the molecular biology laboratory. Enzymes are proteins (encoded by genes) that catalyze a biochemical reaction. In other words, enzymes make biochemical reactions occur much more quickly than they might (or might not) spontaneously.

Enzymes carry on the business of life, whether that be digesting the proteins in the food we eat into the amino acid building blocks that our cells can use, making more DNA, or carrying oxygen molecules along the necessary path so that our cells can use that oxygen. There are countless other examples.

We have learned how to purify enzymes from natural sources and to use them as tools in the laboratory. Some of the most commonly used enzymatic tools in the molecular biology research laboratory and molecular pathology diagnostic laboratory are restriction endonucleases, DNA polymerases, and reverse transcriptase (notice enzyme names always end with the suffix "ase"). When we mix purified DNA or RNA, under controlled conditions, with different enzymes and the necessary ingredients chosen to accomplish a specific task, we are manipulating DNA or RNA to learn more about it, interrogate it in specific ways, and find any clues (research laboratory) or known markers (diagnostics laboratory) of disease.

EST

EST stands for Eastern Standard Time, but with respect to this book we're talking about "expressed sequence tags."

ESTs are usually several hundred base pairs in length. By partially sequencing short stretches of cDNA in cDNA libraries [☞ "cDNA" and "Library"] one can "characterize" expressed genes, for example, characterizing a gene as (1) expressed by human liver, or (2) stomach cancer-specific, or (3) largely homologous to a gene sequence present in GenBank [☞ "GenBank"].

Using ESTs is a handy way for researchers to find expressed genes during a particular snapshot in time, for example, when a model animal is treated with a certain drug. Such research has profound medical and economic implications. With almost 50 million ESTs identified in public databases, ESTs have been instrumental in transcript identification, microarray assembly (generation of appropriate probes), and gene discovery. If you think about the math (~23,000 genes and 50 million ESTs) then you quickly realize that many genes and their transcripts are redundantly represented by the millions of ESTs now known. Remember, ESTs are tags and multiple tags have been identified for each gene and its transcript(s).

Ethics

We want to bring up the idea of ethics as it pertains to issues surrounding DNA technology. For example, *BRCA1*, HD, Dolly, cloning humans, etc.

BRCA1: a gene involved in hereditary breast cancer. Do positive results generated by testing for breast-cancer-causing mutations in this gene have negative implications for a person's protections against potential genetic discrimination by employers and insurers? Many states have passed legislation to protect citizens against these sorts of problems. Federal employees have been protected since the issuance of an Executive Order in 2000. Legislative protection now exists for every American since the President signed into law the Genetic Information Non-Discrimination Act (GINA) in May 2008. Thus, one can now feel safe in gaining this information if certain risk factors (ethnicity, family history) exist.

What is the value of a negative result for breast-cancer-causing mutations? Should those with a positive result for a mutation in this gene get a prophylactic double mastectomy? Should surgery be considered when the disease may not strike until age 80 or 90? Complicated questions all.

HD stands for Huntington's disease, which as we write this in spring 2008, remains incurable. There are definite ethical implications to testing for an incurable disease, regardless of the test result. If the test is negative, there is happiness for that individual but potential survival guilt if a sibling, for example, is not so fortunate. If the test is positive, depression and suicide may follow.

Dolly taught us that we can clone sheep. Should we clone humans? Probably not. Should we practice therapeutic cloning, so intimately associated with stem cells and somatic cell nuclear transfer? Many scientists, physicians, healthcare advocates, disease "interest groups," and others think it's advisable and presents no ethical dilemma. Many others believe the opposite; it is a definite ethical quagmire in need of serious discussion, debate, compromise, and regulation.

Internet sites you can visit to learn more about these issues include the following:

- http://bioethics.gov
- http://www.lbl.gov/Education/ELSI/ELSI.html
- http://www.genome.gov/PolicyEthics/

Ethidium bromide

A commonly used dye; ethidium bromide (EtBr) is a chemical whose structure contains, in part, a hexagonal carbon ring. Imagine this ring structure in a flattened plane; as such it can insert itself (the technical term is "intercalate") between the bases that make up the DNA double helix. Once inserted, EtBr changes the physical characteristics of DNA such that when EtBr-stained DNA is illuminated with ultraviolet light, it fluoresces. Fluorescent, EtBr-stained DNA is easily detected and amenable to photography or imaging so that a permanent record can be made (Figure 16). With this explanation in mind, it is not surprising to learn that EtBr is a mutagen and is something that must be handled with respect and disposed of carefully in the laboratory.

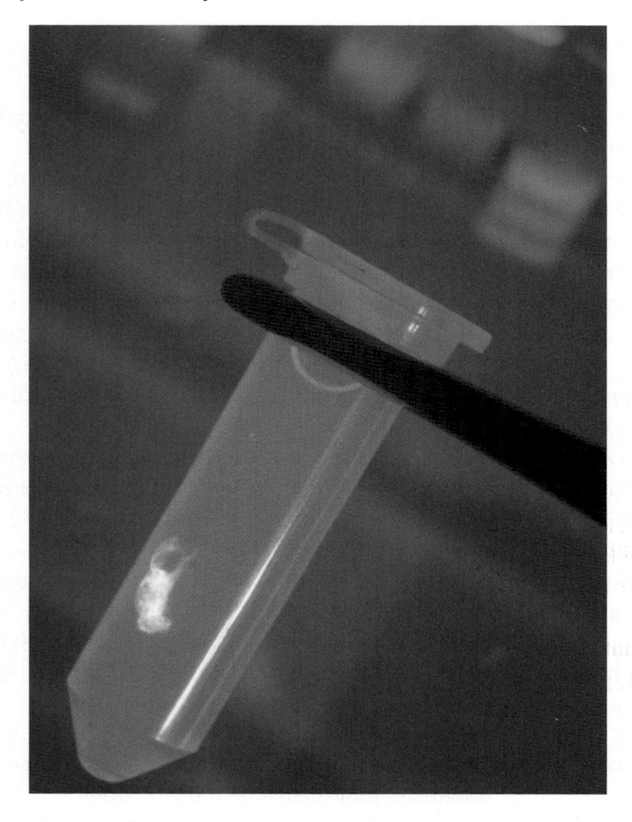

Figure 16. The tube in the foreground contains a white, fluffy, flocculent precipitate near the bottom, which is human DNA. The liquid in the tube is largely alcohol, which causes DNA to precipitate or come out of solution. The reason for the pinkish, reddish hue to the tube is that it is being illuminated by ultra-violet light. In the background of the photo one can see several fluorescent bands. These are pieces of DNA with specific masses or molecular weights. The bands near the top of the photo are heavier than those at the bottom. This photo illustrates DNA migrating based on its molecular weight after electrophoresis in an agarose gel. The fact that the bands are visible is due to the inclusion of ethidium bromide in the gel which causes the DNA to fluoresce when illuminated with ultra-violet light.

Evolution

In past editions of this book, jokes have been made about Creationism in order to denigrate it and to amplify the belief that evolution can explain all life on this planet. This was dangerous practice. Teaching is not achieved through derision but rather through education. When scientists dismiss the thoughts of others as ignorant, scientists invite criticism and mistrust and unconstructive dialog ensues. The scientist's responsibility includes teaching not through disdain, but through observation of facts, data analysis, and experience. Describing "evolution" is beyond the scope of this book except to say that the origin of evolutionary changes in a species begins in the DNA.

For a more intensive review of "evolution" suitable for almost any age, we highly encourage readers to view this address—http://www.pbs.org/wgbh/evolution/—on the Internet. Lastly, we won't argue here against creationism but rather point out what J. Michael Bishop says in his book *How to Win the Nobel Prize* (Harvard University Press, 2003): i.e., Pope John Paul II has stated that evolution is "more than just a theory [and has] proved true." (Source: "The Pope's Message on Evolution and Four Commentaries," *Quarterly Review of Biology* 1997;72:382–383).

Exon [☞ "Expression"]

No, we didn't misspell the name of the oil company (but if you want to use that little device to remember this, that's OK with us; we've got a better memory device for this just below—keep reading). Genes are made up of DNA that is transcribed into messenger RNA (mRNA) and then translated into proteins that carry on the "business of life." The DNA in a gene is arranged in a section-like fashion. Alternating stretches of DNA do and do not code for the ultimate gene product, the protein. The sections of the gene that are ultimately translated into a part of the protein are called exons; **ex**ons are **ex**pressed. The **int**ervening stretches of DNA in between exons are called **int**rons; they are spliced out of the gene when it is made into RNA and serve as regulatory parts of the DNA, or punctuation marks. Some scientists have referred to introns as junk DNA, a term falling increasingly into disfavor as we learn more about DNA, gene structure, and splicing (see Figure 17) [☞ "Intron" and "Splicing"].

Expression

One of us (DHF) once had a discussion with a friend, Brooks Gardner, a professional actor. Brooks is sensitive to the idea of one's "expressing oneself." He found it interesting and comical, in an oxymoronic sort of way, that those who work with DNA speak of gene expression and just what did that mean anyway? Well, here goes:

You may have heard the cliché, "DNA is the stuff of life." What that means is this: our genetic information flows from parent to child through the heritable characteristics embedded in DNA. The business of carrying on life (cellular biochemistry) in each cell, organ, and organ system in our bodies is carried out by proteins. Proteins are responsible for the color of our eyes, for metabolizing medications, for ensuring that the oxygen we breathe is transported to the appropriate place in the cell for utilization, for the elasticity of our skin, for transporting and digesting nutrients, for our immune response to the cold virus our kid brought home from kindergarten, etc. Everything that our bodies do to carry on life is mediated through the action of proteins. Individual proteins are invisible (to the naked eye) structures in our bodies that are generated by "machinery" inside our cells.

The machinery is composed of still more proteins that carry on the business of "protein synthesis." Proteins are synthesized in the gelatinous goop (pretty scientific, huh?) inside

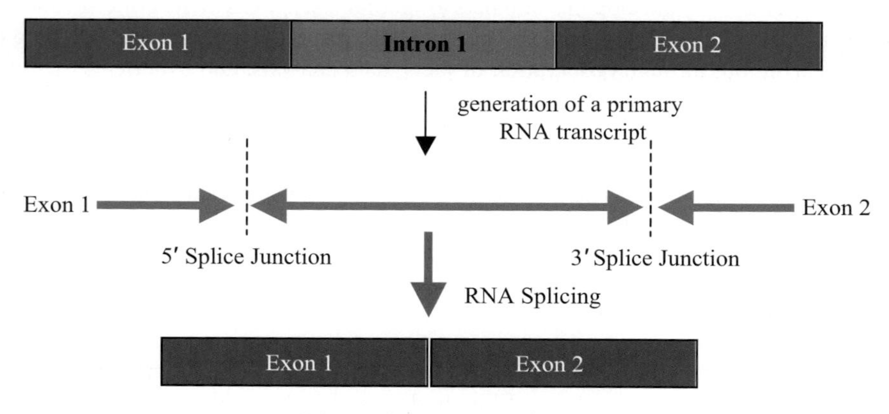

Figure 17. A hypothetical gene is made up of two exons and one intron. If the protein for which this gene codes is needed, under appropriate cellular conditions the gene is faithfully transcribed into a form of RNA called a primary RNA transcript. The primary RNA transcript contains the sequences for the exons and the intron. The ribonucleotide sequences that span the junction of the intron and exons are called splice site sequences. Specialized molecules in the cell recognize these splice sites and snip out the intron sequence from between the two exons, leading to a mature RNA transcript. This mature mRNA can then be translated into a mature protein. That protein contains amino acids coded for by the original exons. It turns out the human genome is a bit fluid, and splice site sequences may vary by a few bases here and there from time to time, cell to cell, and condition to condition, so that slightly different proteins are made when appropriate. This is one way in which 30,000 human genes contain the flexibility to code for so many more proteins.

Modified with permission from Tsongalis GT, Coleman WB. Molecular diagnostics: a training and study guide. Washington, DC: AACC Press, 2002.

our cells called the cytoplasm, which surrounds the cell's nucleus (where the DNA lives). Proteins are made of long stretches of fundamental molecules called amino acids and a protein might count between a few to many thousands of amino acids in its number. How does the protein-synthesizing machinery of the cell's cytoplasm know to assemble, in the correct order, a given set of amino acids to form protein X, Y, or Z? The answer lies in the RNA.

RNA is the intermediary between proteins and DNA. Based on the sequence (of bases) in the DNA, an RNA transcript (loosely speaking, a transcript is a "copy") is generated. That RNA copy contains the instructions given to it by the DNA when the RNA was made from the DNA template. Those instructions are faithfully read by the protein-synthesizing machinery of the cell. (One way to cause mutations is to read those instructions **un**faithfully.) Reading those instructions means translating the code in the RNA from bases (the building blocks of DNA and RNA) to amino acids (the building blocks of proteins).

In summary then, the flow of genetic information is as follows:

$$\text{DNA} \rightarrow \text{RNA} \rightarrow \text{Protein}$$

The bases in the DNA are transcribed into an RNA intermediary (in the cell's nucleus) whose bases are translated into amino acids and ultimately proteins (in the cell's cytoplasm);

those proteins then reside in or on the cell, or are secreted extracellularly, to carry on the business of life. It's worth noting right here (☞ "Tissue-specific gene expression") that while all genes are present in DNA, only stomach-specific gene products are expressed in stomach cells; antibodies, which are proteins that fight infection, are only expressed in immune cells in the blood, etc.

And as so often is the case when you're trying to learn something, there is an exception. For a discussion of that, ☞ "Retroviruses" and "Reverse transcriptase" [☞ "Central dogma of molecular biology;" "Anticodon;" "Ribosomal RNA;" "Tissue-specific gene expression"].

Expression profiling

The FBI uses profiling to catch serial killers. Marketing professionals use profiling to determine which customers are most likely to use a particular product. The Transportation Safety Administration does not use profiling to determine who really ought to be searched at the airport security line. Profiling employs methods to identify unique or telling traits that directly distinguish individuals from the general population. Expression profiling exploits this same concept.

Go back one entry and read "Expression;" note that the flow of information is from DNA → RNA → Protein. Note also that proteins are responsible for everything our bodies do to carry on life, and that not all cells express the same proteins. These expression patterns can be used to understand the basis of cellular differentiation; we wouldn't want a nerve cell doing the job of the liver or vice versa.

Expression profiling compares the gene expression profile of two different populations of cells, e.g., normal to tumor; non-diseased to diseased; no drug to drug-induced. The expression level, that is, the RNA transcripts, from thousands of genes within these cells are then compared and analyzed using sophisticated computer programs that are part of a growing field called bioinformatics. This work is possible due to the advent of microarray technology [☞ "DNA chips"].

Using this approach, research scientists and, increasingly, diagnostic laboratories, are able to identify which genes are expressed under which conditions. For example, the genes that are known to be involved in cellular proliferation and that are aberrantly "up-regulated" in tumor tissue but not normal tissue might be a diagnostic "panel" of genes for a particular kind of cancer. Increasingly, pharmaceutical companies are using this approach to detect abnormal expression patterns against which highly specific drugs may be designed.

Extension

What you ask for on April 14 when your taxes are nowhere near done. With respect to nucleic acids, extension refers to elongation of the growing chain being synthesized, using the parent DNA strand as the template, that is, the daughter strand or RNA transcript.

This is a natural process that occurs during DNA replication or RNA transcription. Elongation is also part of a key molecular diagnostics laboratory process called polymerase chain reaction that employs different reagents and polymerase enzymes (proteins whose job is to synthesize new strands of DNA or RNA) to artificially create new DNA, something akin to making a haystack full of needles [☞ "PCR"].

Extraction controls

Extraction and purification of nucleic acids from cells, always the first step in a molecular diagnostic laboratory test, involve several steps and generally result in highly purified DNA or RNA.

Sometimes, however, things go awry and the nucleic acid is lost during the process. To ensure that useable nucleic acid is recovered from extracted samples, an extraction control is used to guard against false negative results. (Realize that without such a control of the process, a negative result in the test could be a true negative result or could be a false negative result because in fact, DNA or RNA failed to be successfully purified and we were testing material devoid of the target. This is just one example of the sort of rigor clinical laboratory professionals and pathologists apply to the practice of clinical laboratory medicine.)

The extraction control is a separate nucleic acid sequence added to the patient sample, for example, blood, prior to the extraction process. At the end of the laboratory test, the control is also analyzed along with the patient's nucleic acid. If the control nucleic acid is not detected, then something has gone wrong and one should be suspicious of a negative test result for the patient; it may indeed be a false negative result and therefore inappropriate to report to the ordering physician. When an external control is added and recovered, we know the extraction process was successful and we can trust a negative result in the sample as a true negative. Extraction controls are generally used in pathogen (bacteria, viruses, fungi) detection assays [☞ "DNA/RNA extraction and purification"].

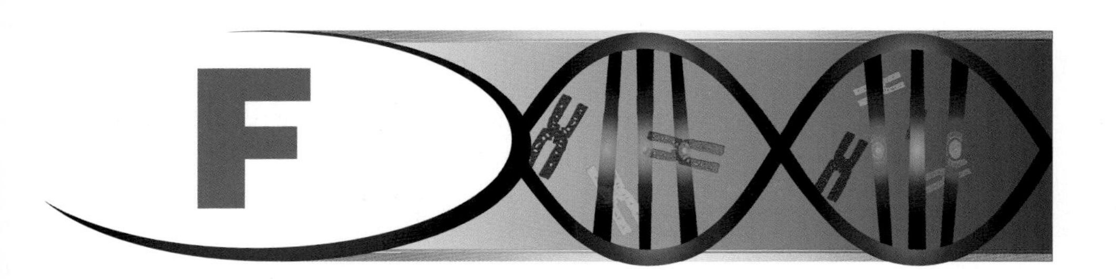

FDA

Food and Drug Administration; the FDA (www.fda.gov) is the U.S. federal agency responsible for overseeing the safety and efficacy of, among other things (like prescription medications), in vitro diagnostic tests and devices (IVDs). The list of FDA-reviewed molecular diagnostics IVDs follows. The list, available electronically at www.amp.org, is reproduced below as Table 3.

Table 3. FDA-Cleared/Approved Molecular Diagnostics Tests (through February 2008)

The following set of tables is a listing of the in vitro molecular diagnostic products that are cleared or approved for diagnostic use in the United States by the Food and Drug Administration (FDA). Such tests are classified as "Medical Devices." More information is available at the website for FDA's Office of In Vitro Diagnostic (OIVD) Evaluation and Safety, *http://www.fda.gov/cdrh/oivd/index.html.* This list is for informational purposes only and is not intended as an endorsement or recommendation in any way of the products and manufacturers listed. Companies are listed alphabetically within each category.

Abbreviations:

ASPE: Allele Specific Primer Extension

510(k): Premarket Notification

bDNA: Branched Chain DNA Signal Amplification

BLA: Biological License Application

DKA: Dual Kinetic Assay

ELISA: Enzyme-Linked Immunosorbent Assay

FISH: Fluorescent *in situ* Hybridization

HPA: Hybridization Protection Assay

IVDMIA, In vitro Diagnostic Multivariate Index Analysis

NASBA: Nucleic Acid Sequence Based Amplification

PCR: Polymerase Chain Reaction

PMA: Premarket Approval

QC: Quality Control

REG: Regulatory Route for Pre-Market Review

RT-PCR: Reverse-Transcription PCR

SDA: Strand Displacement Amplification

TC: Target Capture

TMA: Transcription-Mediated Amplification

Infectious Disease Tests—Bacterial

Test	Manufacturer	Test Name	Method	Reg
Bacillus anthracis	Idaho Technology, Inc., Salt Lake City, UT	Joint Biological Agent Identification and Diagnostic System (JBAIDS) Anthrax Detection kit	Real-time PCR	510(k)
Candida albicans	AdvanDx, Woburn, MA	*C. albicans* PNA FISH	PNA FISH	510(k)
Chlamydia trachomatis detection (single organism)	Qiagen, Germantown, MD	HC2® CT ID	Hybrid Capture	510(k)
	Gen-Probe, Inc., San Diego, CA	APTIMA CT® Assay	TC, TMA, DKA	510(k)
	Gen-Probe, Inc.	PACE® 2 CT Probe Competition Assay (Ct-confirmation test)	HPA	510(k)
	Roche Molecular Diagnostics, Pleasanton, CA	AMPLICOR® CT/NG Test for *C. trachomatis*	PCR	510(k)
	Roche Molecular Diagnostics	COBAS AMPLICOR® CT/NG Test for *C. trachomatis*[1]	PCR	510(k)
Neisseria gonorrhoeae detection (single organism)	Qiagen	HC2® GC ID	Hybrid Capture	510(k)
	Gen-Probe, Inc.	APTIMA® GC Assay	TC, TMA, DKA	510(k)
	Gen-Probe, Inc.	PACE® 2 GC Probe Competition Assay (GC-confirmation test)	HPA	510(k)
	Roche Molecular Diagnostics	AMPLICOR® CT/NG Test for *N. gonorrhoeae*	PCR	510(k)
	Roche Molecular Diagnostics	COBAS AMPLICOR® CT/NG Test for *N. gonorrhoeae*[1]	PCR	510(k)
C. trachomatis and *N. gonorrhoeae* detection	Becton, Dickinson & Company, Franklin Lakes, NJ	BD ProbeTec™ ET *C. trachomatis* and *N. gonorrhoeae* amplified DNA Assay	SDA	510(k)
	Qiagen	HC2® CT/GC Combo Test	Hybrid Capture	510(k)
	Gen-Probe, Inc.	APTIMA Combo 2® Assay	TC, TMA, DKA	510(k)
	Gen-Probe, Inc.	PACE® 2C CT/GC	HPA	510(k)
	Roche Molecular Diagnostics	AMPLICOR® CT/NG Test	PCR	510(k)
	Roche Molecular Diagnostics	COBAS AMPLICOR™ CT/NG Test[1]	PCR	510(k)
Enterococcus faecalis	AdvanDx	*E.* faecalis PNA FISH	PNA FISH	510(k)

Francisella tularensis	Idaho Technology, Inc.	Joint Biological Agent Identification and Diagnostic System (JBAIDS) Tularemia Detection kit	Real-time PCR	510(k)
Gardnerella, Trichomonas vaginalis, and *Candida spp.* Detection	Becton, Dickinson & Company	BD Affirm™ VPIII Microbial Identification Test	Hybridization	510(k)
Group A *Streptococci* detection	Gen-Probe, Inc.	Group A Strep direct (GASD)	HPA	510(k)
Group B *Streptococci* detection	Cepheid, Sunnyvale, CA	Smart GBS	Real-time PCR	510(k)
	Cepheid	Xpert™ GBS	Real-time PCR	510(k)
	Gen-Probe, Inc.	Group B AccuProbe®	HPA	510(k)
	Becton, Dickinson & Company	IDI-Strep B™ Assay	Real-time PCR	510(k)
Legionella pneumophilia detection	Becton, Dickinson & Company	BD ProbeTec™ ET *L. pneumophilia* amplified DNA Assay	SDA	510(k)
MRSA for *Staphylococcus aureus*—Screening assay	Becton, Dickinson & Company	IDI-MRSA™ Assay	Real-time PCR	510(k)
	Cepheid	Xpert™ MRSA	Real-time PCR	510(k)
MRSA for S. *aureus*—Diagnostic assay	Becton, Dickinson & Company	GeneOhm StaphSR	Real-time PCR	510(k)
Mycobacterium tuberculosis detection	Gen-Probe, Inc.	AMPLIFIED™ *M. tuberculosis* Direct Test (MTD)	TMA	PMA
	Roche Molecular Diagnostics	AMPLICOR™ *M. tuberculosis* Test	PCR	PMA
Mycobacteria spp., different fungi and bacteria culture confirmation[2]	Gen-Probe, Inc.	AccProbe® Culture Identification Tests	HPA	510(k)
S. *aureus*	AdvanDx	S. *aureus* PNA FISH	PNA FISH	510(k)

[1]*C. trachomatis* and *N. gonorrhoeae* detection may now be performed using the Roche COBAS Amplicor system directly from Cytyc Corporation's ThinPrep Pap test collection kit; this use is FDA approved.

[2]*Campylobacter spp., Enterococcus spp.*, Group B *Streptococcus, Haemophilus influenzae, Neisseria gonorrhoeae, Streptococcus pneumoniae, Staphylococcus aureus, Listeria monocytogenes*, Group A *Streptococcus, Mycobacterium avium, Mycobacterium intracellulare, Mycobacterium avium* complex, *Mycobacterium gordonae, Mycobacterium tuberculosis* complex, *Mycobacterium kansasii, Blastomyces dermatitidis, Coccidioides immitis, Crytococcus neoformans, Histoplasma capsulatum.*

Infectious Disease Tests—Viral

Test	Manufacturer	Test Name	Method	Reg
Avian Flu	Centers for Disease Control and Prevention, Atlanta, GA	Influenza A/H5	Real-time RT-PCR	510(k)
Cytomegalovirus detection	Qiagen	HC1® CMV DNA Test	Hybrid Capture	510(k)
	bioMerieux, Inc., Durham, NC	CMV pp67 mRNA	NASBA	510(k)
Enterovirus detection	Cepheid	Xpert™ EV	Real-time PCR	510(k)
HCV Qualitative detection	Gen-Probe, Inc.	VERSANT® HCV RNA	TMA	PMA
	Roche Molecular Diagnostics	AMPLICOR™ HCV Test, v2.0	PCR	PMA
	Roche Molecular Diagnostics	COBAS AMPLICOR™ HCV Test, v2.0	PCR	PMA
HCV Quantitation	Siemens Healthcare Diagnostics, Deerfield IL	VERSANT® HCV RNA 3.0 Assay (bDNA)	bDNA	PMA
HIV drug resistance testing	Celera Diagnostics, Alameda, CA	ViroSeq™ HIV-1 Genotyping System	Sequencing	510(k)
	Siemens Healthcare Diagnostics	TruGene™ HIV-1 Genotyping and Open Gene DNA Sequencing System	Sequencing	510(k)
HIV Quantitation	Abbott Molecular, Inc., Des Plaines, IL	Abbott Real-time HIV-1	Real-time RT-PCR	PMA
	Siemens Healthcare Diagnostics	VERSANT® HIV-1 RNA 3.0 Assay (bDNA)	bDNA	PMA
	bioMerieux, Inc.	NucliSens® HIV-1 QT	NASBA	PMA
	Roche Molecular Diagnostics	AMPLICOR HIV-1 MONITOR™ Test, v1.5	RT-PCR	PMA
	Roche Molecular Diagnostics	COBAS AMPLICOR HIV-1 MONITOR™ Test, v1.5	RT-PCR	PMA
	Roche Molecular Diagnostics	COBAS® AmpliPrep/COBAS® TaqMan HIV-1 Test	RT-PCR	PMA
HBV/HCV/HIV for blood donations	BioLife Plasma Services, L.P., Deerfield, IL	Hepatitis C Virus (HCV) Reverse Transcription (RT) Polymerase Chain Reaction (PCR) assay	RT-PCR	BLA
	BioLife Plasma Services, L.P.	Human Immunodeficiency Virus, Type 1 (HIV-1) Reverse Transcription (RT) Polymerase Chain Reaction (PCR) assay	RT-PCR	BLA

	Gen-Probe, Inc. (distributed by Chiron)	Procleix™ HIV-1/HCV Assay	TMA	510(k)
	Gen-Probe, Inc. (Chiron)	Chiron Procleix™ HIV-1/HCV Controls	QC controls	510(k)
	Gen-Probe, Inc. (Chiron)	Chiron Procleix™ HIV/HCV Proficiency Panel	QC proficiency panel	510(k)
	National Genetics Institute, Los Angeles, CA	UltraQual™ HCV RT-PCR Assay	RT-PCR	BLA
	National Genetics Institute	UltraQual™ HIV-1 RT-PCR Assay	RT-PCR	BLA
	Roche Molecular Diagnostics	COBAS AmpliScreen™ HBV Test	PCR	BLA
	Roche Molecular Diagnostics	COBAS AmpliScreen™ HCV Test, v2.0	RT-PCR	BLA
	Roche Molecular Diagnostics	COBAS AmpliScreen™ HIV-1 Test, v1.5	RT-PCR	BLA
Human Papillomavirus Testing	Qiagen	HC2® HR and LR	Hybrid Capture	PMA
	Qiagen	HC2®HPV HR	Hybrid Capture	PMA
	Qiagen	HC2® DNA with Pap	Hybrid Capture	PMA
Respiratory virus panel	Luminex Molecular Diagnostics, Toronto, Canada	xTAG Respiratory Viral Panel[1]	PCR, ASPE, Tag sorting	*de novo* 510(k)
	Prodesse, Waukesha, WI	ProFlu+™ Assay[2]	Multiplex Real-time PCR	510(k)
West Nile for blood donations	Gen-Probe, Inc. (Chiron)	Procleix WNV	Real-time PCR	510(k)
	Roche Molecular Diagnostics	Cobas TaqScreen WNV	PCR	510(k)

[1]Viruses include three types of Influenza A; a strain of influenza B; adenovirus; respiratory syncytial viruses A and B; Metapneumovirus; Parainfluenzas 1, 2, and 3; and rhinovirus.

[2]Viruses include influenza A and B viruses and respiratory syncytial virus.

Molecular Diagnostic Tests—Human

Test	Manufacturer	Test Name	Method	Reg
B-Cell Chronic Lymphocytic Leukemia (B-CLL)	Abbott Molecular	CEP®12 DNA Probe Kit	FISH	510(k)
Breast Cancer: determination of likelihood of metastasis	Agendia, Amsterdam, Holland	MammaPrint	Microarray analysis	510(k); IVDMIA
	Veridex, LLC, Warren, NJ	GeneSearch™ BLN Test	Real-time RT-PCR	PMA
Breast Cancer: detection of amplifications and deletions of the *TOP2A* gene	Dako Denmark A/S, Glostrup, Denmark	*TOP2A* FISH pharmDx™ Kit	FISH on FFPE breast tissue	510(k)
Chromosome 8 Enumeration (CML, AML, MPD, MDS)	Abbott Molecular	CEP®8 DNA Probe Kit	FISH	510(k)
Cystic Fibrosis	Luminex Molecular Diagnostics	Tag-It™ Mutation Detection Kit CFTR 40+4	PCR, ASPE, Tag sorting	*de novo* 510(k)
	Osmetech Molecular Diagnostics, Pasadena, CA	eSensor® Cystic Fibrosis Carrier Detection System	PCR, Probe Hybridization	510(k)
Drug-Metabolizing Enzymes	Roche Molecular Diagnostics	AmpliChip™ Cytochrome P450 Genotyping Test	Microarray	510(k)
	Nanosphere, Inc., Northbrook, IL	Verigene® Warfarin Metabolism Nucleic Acid Test	Multiplex Gold Nanoparticle Probes	510(k)
Factor II (prothrombin)	AutoGenomics, Inc., Carlsbad, CA	INFINITI™ System Assay for Factor II	PCR and Detection Primer Extension	510(k)
	Roche Molecular Diagnostics	Factor II (prothrombin) G20210A kit	Real-time PCR	510(k)
	Nanosphere, Inc.	Verigene® F2 Nucleic Acid Test	Multiplex Gold Nanoparticle Probes	510(k)
Factor V Leiden	AutoGenomics, Inc.	INFINITI™ System Assay for Factor V	PCR and Detection Primer Extension	510(k)
	Roche Molecular Diagnostics	Factor V Leiden kit	Real-time PCR	510(k)
	Nanosphere, Inc.	Verigene® F5 Nucleic Acid Test	Multiplex Gold Nanoparticle Probes	510(k)
Factor II (Prothrombin) and Factor V Leiden	AutoGenomics, Inc.	INFINITI™ System Assay for Factor II & Factor V	PCR and Detection Primer Extension	510(k)

	Nanosphere, Inc.	Verigene® F5/F2 Nucleic Acid Test	Multiplex Gold Nanoparticle Probes	510(k)
Factor II (Prothrombin), Factor V Leiden and MTHFR	Nanosphere, Inc.	Verigene® F5/F2 MTHFR Nucleic Acid Test	Multiplex Gold Nanoparticle Probes	510(k)
HLA Typing	Biotest Diagnostics Corp., Denville, NJ	Biotest HLA SSP	PCR	510(k)
	Biotest Diagnostics Corp.	Biotest DQB-ELPHA	Enzyme linked DNA Probe Hybridization	510(k)
	Invitrogen, Carlsbad, CA	Dynal Reli SSO HLA-A Typing Kit	PCR, Probe Hybridization	510(k)
	Invitrogen	Dynal Reli SSO HLA-B Typing Kit	PCR, Probe Hybridization	510(k)
	Invitrogen	Dynal Reli SSO HLA-Cw Typing Kit	PCR, Probe Hybridization	510(k)
	Invitrogen	Dynal Reli SSO HLA-DQB1 Typing Kit	PCR, Probe Hybridization	510(k)
	Invitrogen	Dynal Reli SSO HLA-DRB3/4/5 Typing Kit	PCR, Probe Hybridization	510(k)
	GTI, Brookfield, WI	GTI PAT HPA-1 (P1) Genotyping kit	PCR, ELISA	510(k)
	One Lambda, Inc., Canoga Park, CA	Micro SSP HLA Class II DNA Typing Kit	PCR	510(k)
	Invitrogen	Pel Freez HLA High Resolution SSP UniTray	PCR	510(k)
HER-2 Status	Abbott Molecular	PathVysion®	FISH	PMA
Initial diagnosis of bladder cancer in patients with hematuria and monitoring tumor recurrence of bladder cancer	Abbott Molecular	UroVysion®	FISH	PMA
5,10-methylenetetra hydrofolate reductase (MTHFR)	Nanosphere, Inc.	Verigene® MTHFR Nucleic Acid Test	Multiplex Gold Nanoparticle Probes	510(k)
Prenatal (Chromosome 13, 18, 21, X & Y)	Abbott Molecular	AneuVysion®	FISH	510(k)
Sex Mismatched Bone-Marrow Transplantation	Abbott Molecular	CEP®X/Y DNA Probe Kit	FISH	510(k)

Molecular Diagnostic Controls

Type of Control	Manufacturer	Control Name	Approved Use	Reg
C. trachomatis and *N. gonorrhoeae* Controls (positive and negative controls)	Bio-Rad Laboratories, Irvine, CA	Amplichek CT/GC Controls	For diagnostic test kits that detect *C. trachomatis* and *N. gonorrhoeae* from swabs or urine	510(k)
C. trachomatis and *N. gonorrhoeae* Positive Quality Controls	SeraCare Life Sciences, West Bridgewater, MA	ACCURUN 341 *C. trachomatis/N.* gonnorrhoeae DNA Positive Control	For diagnostic test kits that detect *C. trachomatis* and *N. gonorrhoeae* from swabs or urine	Pending 510(k)
CMV DNA Controls	SeraCare Life Sciences	ACCURUN 350 CMV DNA Positive Quality Control	For in vitro tests that detect CMV DNA	Pending 510(k)
Cytochrome P450 2D6 gene (CYP2D6)	ParagonDx, LLC, Morrisville, NC	CYP2D6 *4A/*2AxN	For diagnostic testing for the CYP450 *2D6* gene	510(k)
	ParagonDx, LLC	CYP2D6 *2M/*17	For diagnostic testing for the CYP450 *2D6* Gene	510(k)
	ParagonDx, LLC	CYP2D6 *29/*2AxN	For diagnostic testing for the CYP450 *2D6* gene	510(k)
	ParagonDx, LLC	CYP2D6 *6B/*41	For diagnostic testing for the CYP450 *2D6* gene	510(k)
	ParagonDx, LLC	CYP2D6 *1/*5	For diagnostic testing for the CYP450 *2D6* gene	510(k)
	ParagonDx, LLC	CYP2D6 *3A/*4A	For diagnostic testing for the CYP450 *2D6* gene	510(k)
HBV Controls	AcroMetrix, Benicia, CA	OptiQual™ HBV DNA Positive Control	For diagnostic test kits that detect HBV DNA	Class I exempt
	AcroMetrix	VeriSure Pro HBV	Testing in donors	510(k)
	SeraCare Life Sciences	ACCURUN 325 Hepatitis B Virus DNA Positive Control	For diagnostic test kits that detect HBV DNA	Pending 510(k)
HCV Control	AcroMetrix	OptiQual™ HCV RNA Positive Control	For diagnostic test kits that detect HCV RNA	Class I exempt
	AcroMetrix	VeriSure Pro HCV	Testing in donors	510(k)

	SeraCare Life Sciences	ACCURUN 305 HCV RNA Positive Control	For diagnostic test kits that detect HCV RNA	510(k)
HIV-1 Control	AcroMetrix	OptiQual™ HIV-1 RNA Positive Control	Control ranges to quantify HIV-1 RNA	Class I exempt
	SeraCare Life Sciences	ACCURUN 315 HIV-1 RNA Positive Control	Control ranges to quantify HIV-1 RNA	510(k)
HIV-1/HCV Controls	AcroMetrix	VeriSure Pro HIV-1	Testing in donors	510(k)
	Gen-Probe, Inc.	Chiron Procleix HIV-1/HCV Controls	Testing in donors	510(k)
	Gen-Probe, Inc.	Chiron Procleix HIV-1/HCV Proficiency Panel	HIV-1/HCV Proficiency Panel	510(k)
HIV-1/HCV/HBV Controls	AcroMetrix	VeriSure Pro Negative	Testing in donors	510(k)
	AcroMetrix	VeriSure Triplex HIV-1 RNA, HCV RNA, HBV DNA	For diagnostic test kits that detect HIV-1 RNA, HCV RNA, and HBV DNA	Pending Class I exempt
	SeraCare Life Sciences	ACCURUN 345 HIV-1 RNA, HCV RNA, HBV DNA Positive Quality Control Series 150	For diagnostic test kits that detect HIV-1 RNA, HCV RNA, and HBV DNA	510(k)
	SeraCare Life Sciences	ACCURUN 803 Nucleic Acid Negative Quality Control (HIV-1, HCV, HBV)	For diagnostic test kits that detect HIV-1 RNA, HCV RNA, and HBV DNA	510(k)
Human Papillomavirus (HPV) DNA Controls	SeraCare Life Sciences	ACCURUN 370 HPV DNA Positive Quality Control	For in vitro tests that detect HPV DNA in human cervical samples	510(k)
	SeraCare Life Sciences	ACCURUN 870 HPV DNA Negative Quality Control	For in vitro tests that detect HPV DNA in human cervical samples	Pending 510(k)
Human Genomic DNA Control	ParagonDx, LLC	Gentrisure™ Human Genomic DNA Reference Control	For quality control of human DNA tests	510(k)
West Nile Virus (WNV) RNA Controls	SeraCare Life Sciences	ACCURUN 365 WNV Positive quality Control	For in vitro tests that detect WNV RNA in human plasma from blood donors	501(k)
	SeraCare Life Sciences	ACCURUN 865 WNV Negative quality Control	For in vitro tests that detect WNV RNA in human plasma from blood donors	501(k)
	AcroMetrix	VeriSure Pro WNV External Controls	For use with the Procleix WNV Assay to detect WNV RNA in human plasma from blood donors	510(k)

Molecular Diagnostic Systems

Type of System	Manufacturer	System Name	Approved Use	Reg
Amplified Molecular Diagnostic Testing Instruments	Abbott Molecular	Abbott m2000™ (m2000sp + m2000rt)	Abbott Real-time HIV-1	PMA
	Becton Dickinson and Co.	BD Viper™ System	*C. trachomatis* & *N. gonorrhoeae*	510(k)
	Gen-Probe, Inc.	TIGRIS® DTS™ System	Gen-Probe's APTIMA Combo 2® Assay & WNV on blood donations	510(k)
	Gen-Probe, Inc.	Chiron Procleix® Semi-Automated Instrument System	Procleix™ HIV-1/HCV Assay	510(k)
	Gen-Probe, Inc.	Procleix® TIGRIS System	Test Donated Blood with the PROCLEIX ULTRIO® Assay & Procleix™ WNV Assay	510(k)
	Roche Molecular Diagnostics	COBAS® AmpliPrep/ COBAS® AMPLICOR	COBAS® AmpliPrep/COBAS® AMPLICOR HBV MONITOR Test, v 2.0	510(k)
	Roche Molecular Diagnostics	COBAS® AmpliPrep/ COBAS® AMPLICOR	COBAS® AmpliPrep/ COBAS® AMPLICOR HCV MONITOR Test, v 2.0	510(k)
	Roche Molecular Diagnostics	COBAS® AmpliPrep/ COBAS® AMPLICOR	COBAS® AmpliPrep/COBAS® AMPLICOR HIV-1 MONITOR Test, v 1.5	510(k)
	Roche Molecular Diagnostics	COBAS® AmpliPrep/ COBAS® TaqMan	COBAS® AmpliPrep/COBAS® TaqMan HBV Test	Pending 510(k)
	Roche Molecular Diagnostics	COBAS® AmpliPrep/ COBAS® TaqMan	COBAS® AmpliPrep/COBAS® TaqMan HCV Test	Pending 510(k)
	Roche Molecular Diagnostics	COBAS® AmpliPrep/ COBAS® TaqMan	COBAS® AmpliPrep/COBAS® TaqMan HIV-1 Test	PMA
	Siemens Healthcare Diagnostics	VERSANT™ 440 Molecular System	VERSANT HCV RNA 3.0 assay	510(k)

Microarray Systems	Affymetrix, Inc., Santa Clara, CA and Roche Molecular Diagnostics, Pleasanton, CA	Affymetrix GCS 3000Dx Instrumentation System	AmpliChip CYP450 Genotyping Test	510(k)
	AutoGenomics, Inc.	INFINITI™ System Assay for Factor II and Factor V	Factor II (Prothrombin) G20210G & Factor V Leiden G1691A	510(k)
	AutoGenomics, Inc.	INFINITI™ 2C9 & VKORC1 Multiplex Assay for Warfarin	CYP450 2C9 & VKORC1 Genotyping Test	510(k)
Real-Time PCR Amplification Systems	Roche Molecular Diagnostics	LightCycler Instrument v. 1.2	Factor II (prothrombin) G20210A kit & Factor V Leiden kit	510(k)
	Roche Molecular Diagnostics	COBAS Taqman™ Analyzer	COBAS AmpliPrep™ System, HIV, HCV, and HBV	510(k)
FISH Scanning Platform	BioView, Ltd., Rehovot, Israel	BioView Duet Scanning System	Peripheral blood and bone marrow (hematological probes), amniotic fluids (x, y, 18, 13, & 21 probes), urine (Vysis UroVysion probe)	510(k)
	Abbott Molecular	Vysis AutoVysion™ System	Vysis PathVysion® HER-2 DNA Probe Kit	510(k)
	Bioview Ltd.	BioView Duet Scanning System	Peripheral blood, bone marrow, and amniotic fluid samples; urine samples using the Vysis Uro Vysion probe	510(k)

Table reproduced, with permission, from the Association for Molecular Pathology (www.amp.org).

FDA-approved/cleared

When a diagnostics company files an application to the FDA for permission to market an in vitro diagnostic product (IVD), the company has two different filing options: (1) FDA approval and (2) FDA clearance. The type of application depends on the IVD classification (Class I, II, or III) and whether or not there is an existing predicate device (a device that can serve as something of a precedent).

If there is not an existing predicate device, the applicant submits a pre-market approval (PMA) application, which undergoes a rigorous review process. Once it passes this process, the kit/test/device is considered FDA-approved.

If a predicate device does exist, a pre-market notification, called a 510(k), application is submitted, and when it passes this process the kit/test/device is called FDA-cleared.

Of course, this is the *Reader's Digest* condensed version of the differences between filing for FDA approval or FDA clearance. Additional factors are involved in determining which regulatory route is followed. There are also additional filing options such as *de novo*, special, and abbreviated 510(k)s.

For more detailed information, visit http://www.fda.gov/cdrh/oivd/regulatoryoverview.html#8. [☞ "Analyte-specific reagents;" "Research use only;" "Home-brew assay"].

Fluorescence resonance energy transfer (FRET)

[☞ "Real-time PCR"]

Fluorescent in situ hybridization (FISH)

[☞ "Cytogenetics"]

Fragile X syndrome

Fragile X syndrome is the most common form of hereditary mental retardation in males. The frequency of the disorder is one in 1,000 to 1,500 individuals.

In addition to mental retardation, Fragile X syndrome is associated with characteristic clinical symptoms including developmental delay, long and prominent ears, high arched palate, prominent jaw, long face, hyperextensible joints, hand calluses, characteristic behavioral difficulties and neurological findings, double-jointed thumbs, single palmar crease, flat feet, macro-orchidism (overly large testicles), and more.

The disorder got its name from the way in which it used to be diagnosed in the clinical laboratory, usually the cytogenetics laboratory [☞ "Cytogenetics"]. A blood specimen was taken from the patient, and the purified white blood cells were grown in the laboratory in such a way that cytogeneticists could actually observe, microscopically, a site on the X chromosome of these cells that could be induced to break, due to presence of a particularly fragile site. We now know that the site is fragile, in affected patients, because of an unusual characteristic of the DNA there.

Unaffected individuals have at this DNA site a series of three nucleotides, cytosine-guanine-guanine, or CGG, that is naturally repeated anywhere from six to 52 times. In affected individuals, that CGG trinucleotide is repeated more than 200 times, sometimes extending into the thousands. This is a destabilizing feature of the DNA and explains why a break could be induced there when these cells were grown in the laboratory. Individuals with between 52 and 200 repeats of CGG are in the carrier category for the disorder and have no symptoms of fragile X syndrome. In the clinical laboratory we also look at special side groups on the DNA, called methyl groups (-CH_3), to give us insight into carrier versus affected status.

This form of mutation, trinucleotide repeat amplification, has been shown to occur in several other diseases, including Huntington's disease and myotonic dystrophy. In fragile X syndrome, the introduction of all those CGGs interferes with normal expression of the gene; it is that lack of expression of an important gene that leads to the clinical symptoms or fragile X phenotype [☞ "Expression" and "Phenotype"].

This DNA abnormality was clarified in the early 1990s, and a direct DNA-based test to detect fragile X syndrome was developed. The DNA test represents a faster, cheaper, and more specific and sensitive way to diagnose the disorder than was previously available through cytogenetic analysis. At the same time it is important to note that the DNA-based fragile X test is highly specific for that disorder, and routine cytogenetic analysis can uncover other abnormalities that would not be detected by molecular fragile X analysis.

Sometimes physicians suspect fragile X syndrome in babies who "fail to thrive." So if some vague non-specific symptoms in a child include fragile X syndrome as well as many other diseases in the differential diagnosis, DNA testing for CGG trinucleotide repeats in the *FMR1* (fragile X mental retardation) gene will only be diagnostic for that disease and rule out no other diseases that may be the cause.

To learn more about fragile X syndrome, talk to your physician or genetic counselor, or contact the National Fragile X Foundation located near San Francisco (925.938.9300; 800.688.8765; http://www.nfxf.org/).

Franklin, Rosalind (1920–1958)

Simply put, it is unlikely that the celebrated DNA double helical structure would have been solved by James Watson and Francis Crick without the X-ray crystallography data of Rosalind Franklin (see Figure 18). One of her X-ray pictures of DNA, number 51

Figure 18. Portrait of Rosalind Franklin.

Reprinted with permission from Science Source/Photo Researchers, Inc. © 2008 Photo Researchers, Inc. All Rights Reserved.

(Figure 19), sparked a "Eureka" moment in Watson and Crick when they saw it and led to their discovery a short time later.

It seems likely that, had she lived, Franklin would have shared in the Nobel Prize awarded to Watson, Crick, and Wilkins (who was the first to obtain clear X-ray diffraction photos of DNA) in 1962 for the discovery of the double helix. Franklin died in 1958 of pneumonia secondary to ovarian cancer; the Nobel Prize is not awarded posthumously. She died with an international reputation as an expert in the research of carbons, coals, and viruses (particularly her work on tobacco mosaic virus).

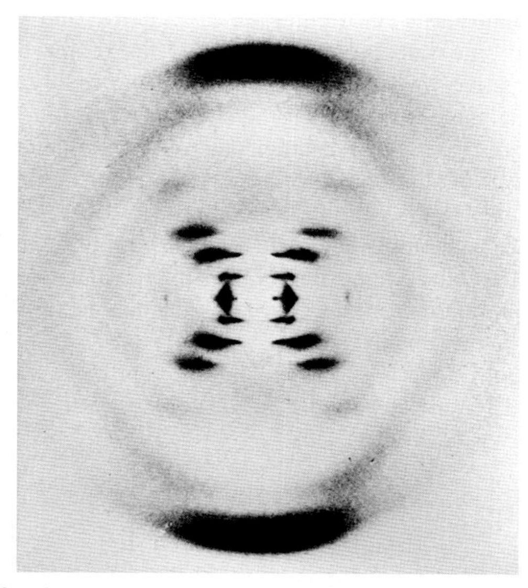

Figure 19. X-ray diffraction photograph of B-DNA taken by Rosalind Franklin in late 1952.

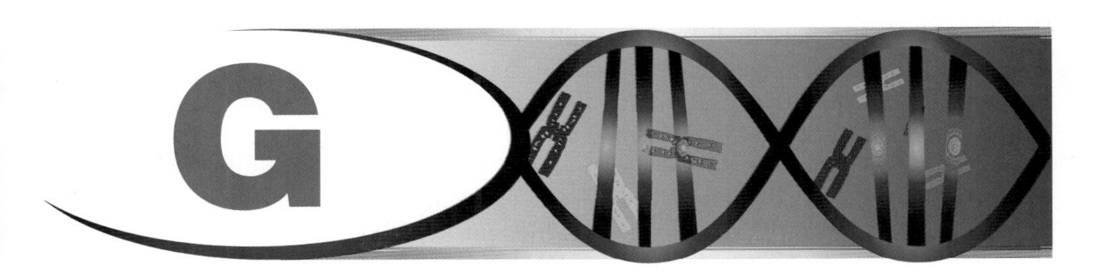

GC-rich

Genes, indeed all DNA, are composed of nucleotides [☞ "Nucleotide"]: guanine (G), cytosine (C), adenine (A), and thymine (T). When a stretch of DNA is particularly high in GC content it is said to be GC-rich.

GC-rich regions of DNA can be troublesome to deal with when using them experimentally or as the target of a DNA diagnostic test. G base-pairs to C in the helix, and A to T. G ≡ C base pairs are bound by three hydrogen bonds while A = T base pairs are bound together by two hydrogen bonds. These facts make GC base pairs more stable, which can contribute to the difficulty of working with GC-rich DNA in the laboratory setting. These difficulties are not insurmountable.

GenBank

GenBank, now over a quarter century old, is the National Institutes of Health (NIH) annotated genetic database comprised of publicly available DNA sequences collected from scientists worldwide. Many tens of billions of bases of sequence information exist at GenBank and the number grows daily.

GenBank is administered and maintained by the National Center for Biotechnology Information (NCBI: http://www.ncbi.nlm.nih.gov/). GenBank is part of the International Nucleotide Sequence Database Collaboration, comprised of the DNA DataBank of Japan (DDBJ; Mishima, Japan); the European Molecular Biology Laboratory (EMBL; Hinxton, UK); and GenBank at NCBI (Bethesda, MD). These organizations exchange data daily.

Access GenBank on the World Wide Web (http://www.ncbi.nlm.nih.gov/Genbank/index.html) [☞ "NCBI"].

Gene [☞ "Expression"]

The most current (as of late 2007) estimates for the number of genes in the human genome range between about 20,500 and about 24,500, small numbers when compared to the ~3 billion base pairs that comprise our chromosomes. A gene is a segment of DNA with a specific architecture (start signals, stop signals, embedded regulatory elements, and more) that the cellular machinery recognizes and transcribes into RNA.

Most genes are eventually translated into proteins that carry out the business of life. Some genes carry out their end function as RNA molecules (for example transfer RNA, or tRNA, molecules).

Genes are responsive to different stimuli. For example, some genes may be turned off until a particular hormone interacts with them to turn them "on" (or "up-regulate" them), so that they can express themselves. It's also a good thing that the genes that code for the proteins that make a stomach cell a stomach cell are not in the "on" position in a liver or brain cell [☞ "Tissue-specific gene expression"]. Environmental insults like cigarette smoke, radiation, or high voltage electric fields cause mutations in some genes. The genes in elementary organisms like bacteria have a different, simpler architecture than genes present in organisms like plants or humans [☞ DNA by the Numbers (#s) section and "Number of genes"].

Gene expression

[☞ "Expression;" "Expression profiling"]

Gene patents

Did you know that the DNA sequence in your body on chromosome six that codes for the *HFE* gene, a gene which when mutated causes a potentially serious but easily treatable disease known as hereditary hemochromatosis, is patented? The same is true for the tumor suppressor gene on chromosome 17, which when mutated is responsible for a large fraction of heritable breast cancer, *BRCA*-1.

B- and T-cell gene rearrangement is a mainstay test of clinical molecular diagnostic laboratories but cannot be practiced without paying a royalty or buying reagents exclusively from a single company. These represent examples of good business practices by the companies who own the patents (patents, it should be emphasized, are an important part of our capitalist economy and provided for by our founders in the Constitution; we all see the benefit of rewarding investment by bestowing limited monopolies called patents). Gene patent holders can control certain aspects of the marketplace and in so doing recoup for their shareholders the investments made in the exploratory science that resulted in these discoveries.

On the other hand, these and similar patents artificially and directly raise the cost of healthcare in the United States. It has been argued that higher costs limit access to excellent healthcare to those who can either afford it or who have superior health insurance. This is a complicated problem with good points on both sides of the argument, and we will not attempt to solve it here. We will leave you, however, with this thought: if you had the skills to draw your own blood and test your own DNA in your own home for a patented gene, you'd be infringing on that patent. You wouldn't owe a royalty, since you wouldn't charge yourself a fee, but there is clearly "something wrong with this picture."

Gene product [☞ "Expression;" "Transcription;" "Translation"]

A gene product is the end result of transcription or translation.

DNA is transcribed into messenger RNA (mRNA), which is then translated into protein. DNA may code for an end product that is RNA, not protein: for example, when DNA codes for the RNA that is one of the building blocks of ribosomes (rRNA) or transfer RNA (tRNA) molecules.

Gene rearrangement [☞ "Chromosomal translocation"]

One of our less scientific colleagues thought of this natural phenomenon and mainstay of molecular diagnostics as gene tampering, which conveys artificiality, something that we in the laboratory barge in and do to the DNA. She wasn't far off except for one key difference: gene rearrangement is natural gene tampering, something the cell does on its own to the DNA it contains.

Gene rearrangement is natural in those species able to mount an immune response, like humans, of course. An important part of our immune response depends upon the production of antibodies (Abs). Even with all the DNA we have in our cells, there is still not enough DNA to encode all the Abs needed to deal with the many antigens (Ags) present in the environment (an Ag is anything that elicits an immune response, like influenza virus, ragweed pollen, or countless other examples). That is because every *successful* immune response depends on the production of unique Abs (more on that below) to *specifically* interact with and help defeat the invading or offending Ag. There are estimated to be only about 20,500–24,500 genes in the human genome; there are many more than 24,500 environmental Ags that can cause an immune response. Clearly our genomes can't afford to devote too large a fraction to immune response; there are other things involved in carrying out the business of life.

Abs are proteins, so their structures are encoded in our DNA. Higher species have evolved a way to deal with the information content or size problem described above. The genes that encode the proteins that compose Abs are arranged in a unique way. They are arranged (and therefore can be *re*arranged) in many different segments or regions; you can think of them as cassettes of DNA coding information. In response to a particular immunological insult, our DNA shuffles around, or rearranges, these cassettes in different ways. There are so many cassettes and so many different ways in which they can be rearranged that unique Abs can be made for specific immunological interaction with an Ag. This so-called gene rearrangement is the way that we generate the necessary Ab diversity to deal with the vast number of Ags in the environment.

We mentioned above that every successful immune response depends on the production of unique Abs; that's not strictly true. There are immune interactions that are cell-mediated as opposed to Ab-mediated. In the same way that our Ab-coding genes rearrange, so, too, do the genes that code for the protein receptors on our immune cells, for example, T cells (a subset of T cells is the target of infection by HIV). These so-called T-cell receptor proteins mediate the interaction between certain Ags and the T cells that help thwart them.

So gene rearrangement is a normal process. We can take advantage of it to diagnose an abnormal process like leukemia or lymphoma. In these diseases, large numbers of a particular clone [☞ definition 2 under "Clone"] of immune cells (B or T cells) have all rearranged an Ab gene or T-cell receptor gene in the same way. Furthermore, this clone of cells is present at an unusually high, disease-causing number. It is the presence of that unique, normal gene rearrangement (think of it as a molecular signature unique for that clone of abnormal, cancerous cells) that can be detected in the molecular pathology laboratory to help in the diagnosis of certain kinds of leukemia or lymphoma. This test is called the B- and T-cell gene rearrangement test and is useful not only in initial diagnosis but also in monitoring the success or failure of therapy, so-called minimal residual disease detection. The test is also useful in determining if the return of disease, should it occur, is due to the same cancerous clone of cells or a different one—information important to the treating oncologist.

While the kind of gene rearrangement described above is normal, sometimes genes are rearranged due to an abnormal event. Chromosomal translocation is such an event. Chromosomal translocation is the abnormal exchange of pieces of chromosomes between each other, for example a piece of chromosome 9 breaking off and attaching to chromosome 22. When this happens, the genes present on the piece of the chromosome that broke off are translocated or moved to a new address inside the nucleus; they can also be thought of as having rearranged, this time as part of an abnormal process. This event occurs in several kinds of cancer and can be detected by a variety of tests available in the clinical molecular pathology laboratory [☞ "Chromosomal translocation"].

Gene testing or DNA-based testing

Clinical and anatomic pathology departments in hospital and reference laboratory settings routinely analyze patient specimens for disease, for example, blood specimens for glucose levels in diabetics, urine specimens to diagnose bacterial infections, or tissue biopsies to determine if cancer is present. In these examples the analyte, or the object for which analysis is being undertaken, is a sugar molecule (glucose); a bacterium (for example, *Chlamydia trachomatis*); and a cancer cell, respectively. In DNA or molecular testing, the analyte is the patient's DNA or sometimes RNA, which is directly analyzed using some of the methods described in this book. Some genetic tests' analytes are gene products, for example, enzymes and other proteins. Some genetic tests are based on microscopic examination of stained or fluorescent chromosomes.

In general, genetic tests refer to diagnostic tests used for things like carrier screening, pre-implantation genetic diagnosis, prenatal diagnostic testing, newborn screening, presymptomatic testing, confirmational diagnosis and (some would argue) identity testing (paternity and forensics). Some use the term "genetic testing" to refer to all DNA testing, including analysis for infection, which is more accurately referred to as "DNA testing." Various types of genetic testing are described below:

- Carrier screening: Testing used to identify individuals who are not affected by a particular disease but who may carry one copy of a disease-causing mutation (when two are required for symptomatic disease). Carrier screening is most useful when a couple is contemplating pregnancy and wants to learn about the risk of passing on a disease that might be prevalent in its family and/or ethnic group.
- Pre-implantation genetic diagnosis: Useful if carrier screening finds a high risk of producing a child who will suffer from serious disease. In this procedure, embryos created in vitro (through in vitro fertilization of a mother's egg with a father's sperm) are analyzed directly for the presence of the mutation. Only those embryos found to be free of the mutation in question would be appropriate choices for implantation into the mother (or a surrogate).
- Prenatal testing: By employing DNA testing methods to search for mutations of interest, an embryo can be assessed for mutations and/or diseases of interest using chorionic villus or amniocytes (obtained during routine OB/GYN procedures) as DNA sources.
- Somatic cell DNA testing: Mutations at the germline level are heritable, by definition. That's because germline cells are those cells involved in propagation of the species, i.e., eggs and sperm. Somatic cells are all the other cells in the body. While there may be a heritable component to many, perhaps all, cancers, the conventional wisdom today

is that most cancers are sporadic; in other words, they occur because of mutations that happened in response to environment, diet, occupational exposure, smoking, etc. We are living longer in Western societies so we see more cancers in older individuals because they have had more "life experience" to accumulate environmentally induced, highly personalized (in other words, not transmissible or heritable), cancer-causing mutations. For example, certain gene "signatures" represented by highly individualized and specific gene rearrangement [☞ "Gene rearrangement"] may be used to determine if a patient's leukemia has recurred or is responding to therapy. This sort of DNA testing is best considered more ordinary and less ethically problematic, if at all, than the sort of heritable testing associated with, for example, pre-symptomatic testing (see below) in the case of Huntington's disease.

- Newborn screening: Thankfully, birth defects are rare. By using screening tests shortly after birth (laboratory tests that can be done at the DNA and/or protein levels) on a drop of dried blood from newborn babies, serious illness associated with some birth defects can be avoided through early diagnosis and therapeutic intervention. All 50 states in the U.S. have had government-mandated newborn screening programs since 1965. There are many fine state and reference laboratories that perform such testing. For example, phenylketonuria (PKU) is a genetic disease that can have extremely serious symptoms if not treated. Fortunately, however, a simple screening test can be done to learn whether or not a baby is affected and therefore whether to implement highly effective dietary therapy (this is why there is a warning message on the side of foods containing aspartame, an artificial sweetener that contains phenylalanine, a normal amino acid that is metabolized well by unaffected individuals but can have drastic consequences in phenylketonurics).
- Pre-symptomatic testing: Genetic testing performed before an individual shows symptoms of a disease. This is useful in cases, for example, where individuals want to learn of their risk of Huntington's disease that may be prevalent in a family (this is a difficult example because this is a fatal, untreatable disease and there are ethical issues aplenty). Another use of this testing is when a family member is being considered as a donor for a relative with end-stage renal disease due to adult-onset polycystic kidney disease. It would be useful to test the potential asymptomatic donor to learn if s/he, too, has the mutations that will ultimately lead to the disease and disqualify him/her from being a kidney donor. Pre-symptomatic testing may also be used to estimate risk of developing diseases such as breast or ovarian cancer (*BRCA1* and *BRCA2* genes) and Alzheimer's disease (*apoE*, *PS1*, *APP* genes).
- Confirmational diagnosis of a symptomatic individual: Examples include testing for the Factor V Leiden mutation in an individual with deep vein thrombosis (blood clotting) so that lifelong blood-thinning therapy may be considered; or testing for certain cystic fibrosis (CF) mutations in infertile males. The genital form of CF is associated with various mutations that cause congenital bilateral absence of the vas deferens and resultant sterility.

Gene therapy

Many diseases are caused by a genetic malfunction. Large amounts of money and effort are being invested into researching the correction of malfunctioning genes. This is known as gene therapy.

Successful gene therapy can take the form of introducing a functional gene into a patient's cells, correcting gene expression and thereby reversing the defect caused by the abnormal gene. Such genetic correction needs to be tissue specific in order to accomplish its task. For example, correcting a cystic-fibrosis-causing mutation will likely need to occur in lung and pancreatic tissue (organs affected by cystic fibrosis) to reverse the improper function (that causes symptoms) in those organs.

In 2002, there was exciting news from a University of Washington, Seattle (UW), research group. Researchers reversed muscle wasting in mice that had been genetically bred to develop muscular dystrophy; they were bred to lack the gene for the protein dystrophin. Mutations in humans that cause lack of or reduced dystrophin expression lead to muscular dystrophy. (This kind of animal experimentation is called a mouse model, where the mouse is used as a "model" for muscular dystrophy.) The UW group showed that the dystrophin gene could be introduced into the muscles of mice that had already begun to show significant disease symptoms and that muscle wasting could be reversed and normal function restored.

Fast forward to late 2007 and the collaborative efforts of Dutch and Belgian physicians and scientists (N Engl J Med 2007;357:2677): The lack of the key protein, dystrophin, in the Duchenne form of Muscular Dystrophy (DMD) is caused by an error in the gene, leading to minimal, if any, dystrophin production. In mouse model systems it was shown that the gene can be "restored" with specific, short stretches of specific nucleotides (oligonucleotides; in this specific case called PRO051) opposite in nature (antisense) to portions of the dystrophin gene, thereby "tricking" the gene into expressing itself, at least partially, leading to dystrophin production. This "trick" or therapy was tried in DMD patients. Intramuscular injection of PRO051 into DMD patients proved safe and, in four patients, restored modest levels of dystrophin production. Other modest success stories exist in diseases like retinal dystrophy, metastatic melanoma (a disease affecting myeloid cells), and others.

Gene therapy can also be defined as introduction of a new function into a cell that is not strictly the introduction of a new gene. For example, cancer cells can be artificially immunostimulated by genetic mechanisms to help "vaccinate" a patient against his or her own tumor.

Vectors are the tools used to deliver therapeutic genetic material into cells. "Vector" is defined as carrier or vehicle. In gene therapy the most common vectors are viruses, taking advantage of the natural role of viruses, which is to infect cells and introduce viral DNA or RNA. There are many associated problems with using potentially pathogenic or deadly viruses for therapy, however. When vectors such as lentiviruses (HIV is in the lentivirus family) are used, much of their genome is modified to remove the dangerous portions, but concern remains. Adenoviruses cause the common cold and are efficient vectors but stimulate the immune system and thus survive in the body a relatively short time. Nonviral vectors are generally poor at transferring their therapeutic DNA to target cells.

The clinical molecular pathology laboratory of the future will not only be a diagnostics laboratory. When gene therapy becomes a reality for treating human disease, the molecular pathology laboratory will be used to identify missing or damaged genes in patients and to find out if these individuals are appropriate candidates for gene therapy. Furthermore, therapeutic agents composed of DNA and RNA will need to be monitored for degradation and purity, although this will likely be a role for the manufacturing pharmaceutical concern. Newly introduced genes in patients will have to be assessed for proper insertion, and for demonstration and quantification of new gene expression. Learn more on the Web from The American Society of Gene Therapy (http://www.asgt.org).

For ethical reasons, gene therapy should only be done on somatic cells; that is, those cells in the body that are not gametes (sperm or eggs). Genetic manipulation of gametes is unethical, for obvious reasons, and is not done by responsible scientists and researchers.

Gene therapy remains a science full of potential but very clearly still in need of especially careful and exhausting research. The death of a gene therapy candidate, Jesse Geisinger, from a rare liver disease (ornithine transcarbamylase deficiency, which prevents the liver from getting rid of the toxic ammonia that accumulates), and the appearance of leukemia in other gene therapy candidates due to the viral vector used to deliver the therapeutic gene have caused great scrutiny in the field. Another death was reported in July 2007 in a trial in which more than 100 similarly treated patients suffered no adverse effects. A redoubling of efforts at safety to recapture public trust in what will clearly be a valuable therapy sometime in the future seems appropriate. As of early 2008, the FDA has approved no human gene therapy products for sale.

GeneTests/GeneClinics

The GeneTests and GeneClinics site (http://www.geneclinics.org/) is free to all who are interested in using it. The site contains information on genetic testing as well as a directory of laboratories that provide testing for genetic disorders. Both research and diagnostic laboratories are included. The site is limited to information about heritable disorders.

Genetic code [☞ "4-3-20"]

A code is a series of items, words, symbols, or the like that make no apparent sense until that code is broken so that it can be read and understood. DNA is written in code. DNA is made up of nucleotides [☞ "Nucleotide"]—guanine (G), cytosine (C), adenine (A), and thymine (T)—which can be thought of as a four-letter alphabet. The combination of those four letters makes up all our DNA, which codes for all our proteins [☞ "Expression" and "Tissue-specific gene expression"].

DNA is composed of four nucleotides and proteins are composed of up to 20 different amino acids. Therefore, it follows logically that one nucleotide *cannot* code for one amino acid. If two-nucleotide combinations (call it a twin) coded for each amino acid, only 16 (4^2) possible twins would exist and that's not enough either for 20 amino acids. Through work with the simple genomes and protein architecture of certain viruses, researchers found that a string of three nucleotides codes for amino acids; these combinations of three are called triplets. If you raise 4 (nucleotides) to the 3rd power, you get 64, and such a coding system can obviously accommodate 20 amino acids.

Work in 1961–1964 by Nirenberg, Matthaei, Ochoa, Khorana, and Leder, which focused on the use of preparing synthetic nucleotides in various orders and detecting which amino acids were produced, led to the remarkable Nobel Prize-winning achievement of cracking the genetic code. Several interesting features were deduced. The triplet code has no punctuation. There is no space or comma between the end of one triplet and the beginning of the next. The start point and reading frame thus become very important, because if something happens that makes decoding proceed from (for example) the normal

ABCABCABC

to the mutant

BCABCABCA,

then all the amino acids produced from that mutant piece of DNA will be in a garbled, incorrect order. There is a need for a precise START triplet codon within the genetic code and, in fact, such a codon exists (AUG). (START means: begin protein synthesis from a particular RNA sequence that will be translated into that protein by the cell.) From this it also follows that the introduction or deletion of one or two bases from a coding sequence is much worse than the introduction or deletion of a triplet. In fact, this is true. So-called out-of-phase or frameshift mutations generally have a worse effect on the resultant mutant protein than mutations that are in phase (insertion or deletion of three nucleotides where only one amino acid is disturbed). In-phase, triplet-based mutations like deletion of a single triplet codon for phenylalanine in the CFTR gene, however, can also have very damaging consequences; in this example, cystic fibrosis. This deletion is in an extremely important functional area of the protein.

The genetic code is universal. All organisms on this planet use the same genetic code. As far as Romulans, Kardassians, Vulcans, and little green men on Mars—well, we just don't know about their genetic codes. Mars-landers, launched by NASA, may someday gather data that yields insights into these questions.

Here's another interesting feature of the code: it is degenerate. There are more combinations of triplets (64) than there are amino acids (20). In fact, several amino acids are encoded by more than one triplet. For example, alanine is encoded by GCU, GCC, GCA, and GCG; valine also has four different triplets that code for it. Only two amino acids have but a single triplet code word. While degenerate, the code is not imperfect. No triplet codes for more than one amino acid.

The first two bases in the triplet codon are more specific (for a given amino acid) than the third base. In the alanine example above, notice that the first two bases are always GC and that any of the other bases found in RNA (U, C, A, or G) can complete the triplet code for alanine. The third base is not as important and tends to "wobble" as Francis Crick, co-discoverer of the double helical nature of DNA, put it. (Remember, RNA is translated into protein and RNA contains U instead of T, as in DNA. That's why there are U bases in the examples above. Don't forget that this whole process is ultimately determined by the base sequence in the master DNA molecule.)

AUG is the triplet that signals the initiation of protein synthesis and happens to code for the amino acid, methionine. Three of the 64 triplets code for no amino acid. UAG, UGA, and UAA signal the cellular machinery to end protein synthesis here (at the STOP codon). The protein is done and so is this entry.

Genetic counseling

Genetic counseling is likely to be a field in need of talented, motivated professionals in the coming decades. As we learn more about DNA, genetics, the human genome, and the full interplay among all those things as they relate to human disease, our need for appropriately trained and certified genetics counselors will continue to grow.

Genetics counselors provide invaluable, confidential information to patients during "one on one" sessions to educate about a disease, a genetic finding in the individual or family, reproductive decisions, etc. Thoughtful, sensitive communication must occur so that patients and families can make intelligent, informed choices that may impact their most personal decisions about health, children, and quality of life.

There are not enough qualified genetics counselors in the United States or the world to deal with the vast amount of genetic data we are collecting today and will use tomorrow. If you are interested in these fields or want to know more about genetic counseling as a career, contact the American Board of Genetic Counseling (http://www.abgc.net) or the National Society of Genetics Counselors in Chicago (www.nsgc.org/).

Genetic engineering

Genetic engineering as a man-made entity is a relatively recent phenomenon. We say that because nature has been engineering new life forms through the manipulation of DNA for eons, a process called evolution. Genetic engineering as a laboratory phenomenon began with the discovery of restriction endonucleases [☞ "Restriction endonucleases"]. We learned how to manipulate DNA in vitro (in the test tube), so that we could recombine it (hence the term, "recombinant DNA") with other pieces of DNA, then insert these so-called clones into bacterial cells and make lots more as the bacteria (or viruses or yeast, which are also suitable cloning vectors) divided [☞ "Clone," definition 1]. Some also refer to this process as gene splicing, although splicing refers to another, more natural phenomenon [☞ "Splicing"].

Genetic engineering has been refined over the few decades of its existence to the point where medically and pharmaceutically important reagents (for example, insulin and interferon, an antiviral drug) are now routinely made on a production manufacturing scale. Genetic engineering is also used in brewing, animal husbandry, fermenting, agriculture, wine making, and other fields. Genetic engineering has contributed significantly to human progress and is a multi-billion dollar industry, spread among many sectors, worldwide.

Genetics Home Reference

This website—http://ghr.nlm.nih.gov—is a great place to learn the basics of genetic disease. Type in a disease name and link to lots of relevant documents written in simple-to-understand language.

Genome [☞ "Human Genome Project"]

The genome of an organism is its complete genetic complement. The genome can also be thought of as the complete set of instructions for reproducing that organism and carrying out its biological function in life—a master blueprint. The DNA in our cells comprises our genome and is contained on chromosomes. When our cells divide, so, too, is the complete genome in those cells duplicated for transmission to each of the resultant daughter cells.

The Human Genome Project successfully completed sequencing all three billion base pairs of the human genome in 2003.

Genomics

The entire DNA sequence, or genome, of many organisms, including *Homo sapiens*, has been completely solved. This plethora of information has necessitated new methods of investigating these genomes for clues relevant to evolution, biological function, diagnostics, therapeutics, etc. Genomics encompasses not only the laboratory-based analytical systems but also computer-based approaches to analyzing individual genomes and comparative analyses among genomes.

The term "genomics" has taken on a broad and generic meaning to encompass all that flows from the sequencing and study of genomes, including but not limited to biotechnology, genomic medicine, personalized medicine, genetic and molecular diagnostics, comparative biology, genetic variation, novel therapeutic solutions, and more. It is the biological equivalent of computer science's Information Technology or IT and in that vein is closely linked to the data analysis portion of genomics, commonly called Bioinformatics. Genomics and bioinformatics will be high growth areas of the economy for decades to come and the opportunities are not relegated to traditional healthcare companies—companies like Google, IBM, Microsoft, HP, and others are making full-fledged endeavors into the field.

Genotype [☞ "Phenotype"]

Genotype refers to unique genetic information contained in an individual organism. Think of it as "gene type." The manifestation of that genotype is known as phenotype.

Eye color is a good example for explaining the difference between genotype and phenotype. Two alleles [☞ "Allele"] or forms of a gene are present in an individual. A person may have inherited one allele for blue eye color from one parent and a second allele for brown eye color from the other parent. Brown eyes are dominant to blue eyes. That means that if you have a blue and a brown allele in your genome, your eyes are brown. Your genotype is that you are heterozygous for brown eye color (a blue plus a brown allele). Your phenotype is brown eye color. The actual manifestation of the genotype is brown.

On the other hand, an individual who has two brown alleles has the same phenotype (brown eyes) as the individual in the above example, but a different genotype. Whereas the first person has one of each allele (heterozygous), the second person has two brown alleles and is said to be homozygous. Two blue eye alleles, of course, lead to the blue eye phenotype.

In terms of disease let's use a common cystic fibrosis (CF)-causing mutation called ΔF_{508}. Delta F 508 refers to the deletion of the amino acid, phenylalanine, normally present at position 508 in a protein, which when mutated (missing), causes CF. Delta is for deletion, F is the abbreviation for phenylalanine, and 508 refers to the relevant position in the protein. Let's consider three individuals in the following example (Table 4):

Table 4. Comparison of Genotype and Phenotype Related to ΔF_{508} Mutation

Individual #	ΔF_{508} Mutation?	Genotype	Phenotype
1	Has no mutated alleles	Normal/normal	Neither a carrier of nor afflicted with cystic fibrosis (CF)*
2	Has one mutated allele	Normal/ΔF_{508}	Carrier for CF but does not have the disease
3	Has two mutated alleles	$\Delta F_{508}/\Delta F_{508}$	Affected with CF

*At least with respect to this mutation; there are hundreds of other mutations that, if present, are known to cause CF. All three individuals have different genotypes. Individuals 1 and 2 have the same phenotype; they are both unaffected with cystic fibrosis, though individual number 2 is a carrier.

Germline [☞ "Somatic cells"]

Unfortunately, unlike "A Chorus Line," a germline is not a musical with microbes kicking up their flagella. Germline cells are the body's reproductive cells, i.e, the sperm and the egg (the scientific term is gamete). Germline cells contain genetic material from the mother and the father that is inherited by the offspring. Since germline DNA is inherited, it is incorporated into the DNA of every cell in the body and passed to subsequent generations. Mutations in germline cells are therefore heritable [☞ "Inheritance"].

Germline cells differ from somatic cells in that somatic cells make up all the rest of the cells in the body (excluding undifferentiated stem cells). Somatic cells are the result of fusion between the sperm and egg. One major difference between germline cells and somatic cells is their number of chromosomes. Germline cells contain 23 single unpaired chromosomes, or one full set of chromosomes, and are referred to as "haploid." Because somatic cells are the result of fusion of two haploid germline cells, they contain 46 single chromosomes, or 23 paired chromosomes, one set of 23 from the mother (egg cell or ovum) and one set of 23 from the father (sperm cell), and are called "diploid" [☞ "Somatic cells"].

Because germline cells contain 23 chromosomes, one set of chromosomes must be "eliminated" during the process of sperm or egg formation. The somatic cells from which germline cells are derived contain pairs of chromosomes, which may or may not contain identical genetic information. If the germline cell gets the chromosome containing a mutant gene, then that gene will be passed to the offspring. It follows, then, that if the germline cell gets the chromosome with the normal gene, then the offspring will not inherit the disease. This is one way genetic diversity occurs. In the molecular diagnostics laboratory, we are able to identify whether or not the offspring of affected parents inherited the normal or the abnormal gene [☞ "Somatic cells"].

GINA

Genetic Information Non-discrimination Act; this act passed Congress overwhelmingly in April 2008 and was signed into law by the President in May 2008. The law protects individuals against discrimination by health insurance companies and employers based on the individual's genetic information. This law of the land not only protects Americans from discrimination due to inherited genetic characteristics, but also is intended to encourage us to take advantage of genetic testing as part of appropriate medical care.

Green fluorescent protein

Green fluorescent protein or GFP, is a protein that glows green when exposed to fluorescent blue light. It was first identified in one species of jellyfish, *Aequorea victoria*, and another form of the protein was later found in the sea pansy. GFP is comprised of 238 amino acids of which a sequence of three (serine-tyrosine-glycine) is responsible for its fluorescence. Proteins that fluoresce blue (BFP), cyan (CFP), and yellow (YFP) have been engineered by mutating these or surrounding residues in GFP.

GFP may be cloned and introduced into a variety of organisms. In 1994 GFP was successfully cloned and expressed in the bacterium *E. coli* (Figure 20) and the nematode *C. elegans*. Subsequently, GFP has been introduced into many bacteria, fungi, plants, flatworms, and mammalian cells including cats, mice, fish, rabbits, and pigs. Cloning of the GFP rabbit Alba for the artist Eduardo Kac for "artistic" purposes in 2000 created much controversy. A GFP pig was mated with a normal pig and gave birth to piglets that exhibited partial fluorescence on their mouths, trotters, and tongues, indicating that the gene can be inherited.

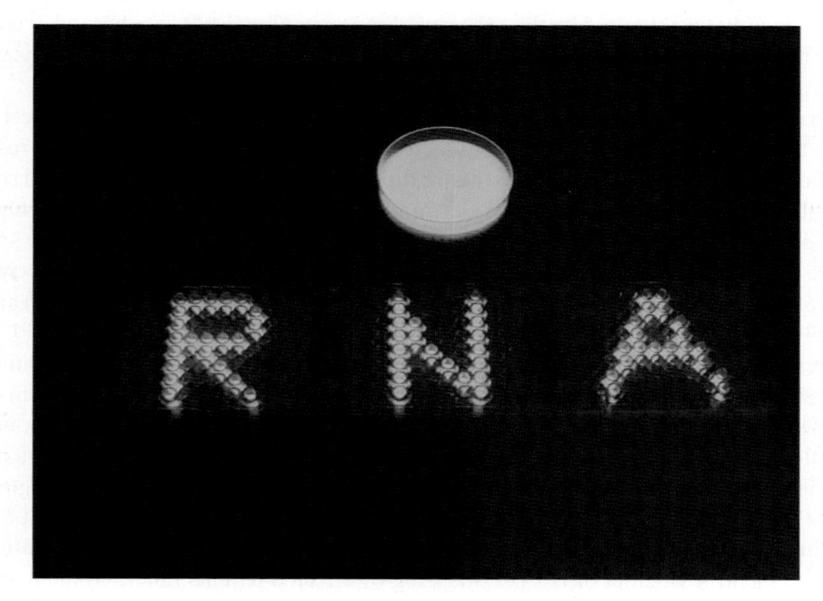

Figure 20. Green fluorescent protein cloned into *E. coli* and visualized using UV light.
Image courtesy of Professor Philip R. Cunningham, Wayne State University, Detroit, Michigan.

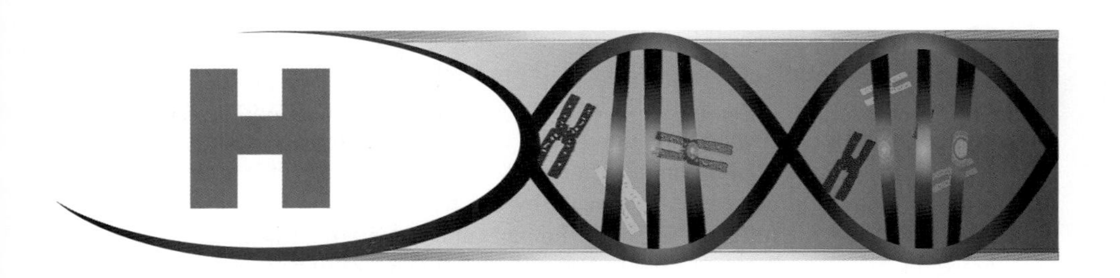

HAC [☞ "BAC;" "YAC"]

Human artificial chromosomes were successfully created in 1997 and are independent replicating chromosomes in human cells.

Cells that contain HACs have an additional chromosome—47 chromosomes per cell instead of the normal 46. They are used as vectors to insert large gene(s) of interest into either human cells or animal models such as the mouse. HACs are smaller than natural chromosomes; however, they contain all the necessary elements required for replication, segregation, and maintenance during and after cell division. Because of these features, HACs are able to survive in human cells for longer periods of time than other gene transfer vehicles.

HACs may be used as vectors for gene therapy to complement genetic deficiencies. Scientists are also creating HACs, which when inserted into stem cells, result in long-term gene expression, producing "cures" for diseases such as cystic fibrosis, diabetes, and certain cancers.

HACs are also used to express human genes in animal model systems to mimic human disease in research and for studies on human chromosome structure, function, and replication.

Hairpins

It's hard to believe this word has relevance in a book about DNA. An explanation is available [☞ "Primer" and "Beacons"].

Haplotype

Somewhat similar to genotype, haplotype is the set of alleles (or single nucleotide polymorphisms (SNPs); [☞ "SNP"]) on one chromosome or part of a chromosome that are linked and usually or often inherited together. If the individual properties on a Monopoly® game board are thought of as individual genes or SNPs, then the entire side of the board that comprises, for example, all the red and yellow properties along with B&O Railroad and other squares is a haplotype.

A clarifying step in relating the now available complete sequence of the human genome to disease is to learn more about haplotypes and search for possible links. Haplotypes that serve as something of a genetic signature for complex diseases like diabetes or cancer may be detectable. Such haplotype signatures may be easier to find and more diagnostic and informative than a particular individual gene or SNP.

Although two individuals may share 99.9% of their DNA, the differences or polymorphisms can be significant with respect to disease risk. Single nucleotide polymorphisms

(SNPs) are the main polymorphisms among individuals. SNPs in close proximity to each other (like the reds and yellows in Monopoly) are inherited from the donating parent, *en masse*, in a block. Such an example of a haplotype may contain several key SNPs that uniquely identify it, and furthermore, potentially associate it with disease.

Finding these unique haplotypes is a goal of the HapMap project [☞ HapMap, next entry] and will simplify the search for disease by one to two orders of magnitude. In other words, approximately 10 million SNPs in the human genome will be simplified to a few hundred thousand unique haplotypes. Other phenotypes that may be studied as a result of completing the HapMap (the complete set of haplotypes in humans)—possible applications that go beyond disease risk—include variation in response to environmental factors, susceptibility to infection with pathogens, and responsiveness (including adverse response) to drugs and vaccines.

HapMap [☞ "Haplotype]

An international effort began in late 2002 to generate an encyclopedia of human haplotypes (and their associations with disease or lack of disease), the so-called HapMap project.

Phase I of the HapMap project, which included the complete data set, was published in late 2005 and Phase II data, which included the analysis of the data set, was published two years later in late 2007. The program's goal was to chart genetic variation within the human genome at an unprecedented level of accuracy. All the data, including its analysis, was placed in the public domain at: http://www.hapmap.org, and is available free of charge.

HER-2/*neu*; HERCEPTIN

The human epidermal growth factor receptor 2 gene or HER-2/*neu* gene is a proto-oncogene [☞ "Proto-oncogene"] which, when mutated, is associated with breast and ovarian cancer (and is also involved in endometrial, gastric, prostate, and kidney cancer).

In the late 1980s, UCLA School of Medicine's Dennis Slamon and his collaborators found that this gene was amplified (present at more than its normal, one diploid copy per cell) in a significant fraction of breast cancer patients. This so-called "HER-2/*neu* gene amplification" changed HER-2/*neu* from a benign proto-oncogene to one that could now be classified as a cancer-causing (or at least cancer-associated) oncogene. HER-2/*neu* gene amplification was found to be associated with decreased overall survival and decreased time to relapse (measured from lumpectomy as time "zero") in those breast cancer patients when compared to patients with only one copy of this gene. Furthermore, Slamon's group showed that the greater the extent of amplification, the shorter these time periods were.

These molecular and clinical observations have been the foundation for development of a specific drug tailored to treat breast cancer patients with amplified levels of HER-2/*neu*. A monoclonal (highly specific) antibody to the protein encoded by HER-2/*neu*, when used in combination with certain chemotherapy regimens, slows tumor progression and increases tumor shrinkage (compared to chemotherapy alone). The antibody is called HERCEPTIN (the technical name is Trastuzumab), is manufactured by Genentech, and is recommended in certain cases of breast cancer where HER-2/*neu* is in fact overexpressed at the protein level. This drug, tailored to work specifically against tumors with a particular gene alteration, is a prime example of pharmacogenomics [☞ "Pharmacogenomics"].

The test used in the clinical laboratory to assess the presence or absence of HER-2/*neu* gene amplification, the result of which is used to determine if a breast cancer patient is an appropriate candidate for HERCEPTIN, is a perfect example of a so-called companion diagnostic test [☞ "Companion diagnostics"]. The U.S. Food and Drug Administration approved HERCEPTIN use in 1998. More information may be obtained online (http://www.herceptin.com).

Heteroplasmy

Heteroplasmy is defined as the existence within an individual of a heterogeneous collection of mitochondria. Mitochondria are cellular organelles [☞ "Organelle"] that contain DNA called mitochondrial DNA (mtDNA). Nuclei are cellular organelles, too, that are more traditionally associated with DNA and called nuclear DNA. All of an individual's mitochondria are derived from one's mother. Some mitochondria may contain disease-causing mutations in their mtDNA. An example of a disease caused by mitochondrial genetic abnormality is Leber Hereditary Optic Neuropathy, which can lead to blindness.

Naturally occurring heteroplasmy may be exploited in forensic investigations to differentiate individuals, for example, suspected perpetrator and crime victim. Heteroplasmy may also be one of two ways in which identical twins may differ genetically; copy number variation is the other way [☞ "Copy number variation"].

Heterozygote [☞ "Homozygote;" "Wild type;" "Allele"] (heterozygous)

An organism is a *heterozygote* if it carries two different forms of a particular gene. This results from inheriting one copy of the gene from one parent and a different copy of the same gene from the other parent.

An example is a person who is a heterozygote for the disease hereditary hemochromatosis type 1 (HH), caused in part by a mutation in the *HFE* gene. In this case, the mutated form of the *HFE* gene is carried on one chromosome, while the normal form of the *HFE* gene is carried in the same position on the other chromosome. Such a heterozygous individual (the genotype) is still generally phenotypically normal.

A heterozygous (for a given gene) individual is termed a carrier (of that mutation) if the person possesses the mutant gene but shows no signs or symptoms of the disease. How can an individual have a mutation and not be affected? This is because of the recessive nature of many mutations [☞ "Inheritance"]. In this case, the normal gene makes enough normal protein to compensate for the mutant or unexpressed protein encoded by the mutant allele. If the heterozygous individual does show disease symptoms, this is known as a dominant mutation; i.e., the abnormal gene produces a protein that is not compensated for by the normal protein.

Other types of heterozygosity also exist. In the above example, the person either did or did not have disease symptoms based on whether the mutation was recessive or dominant. These distinctions are not always clear. In the HH example, the presence of the mutated *HFE* gene is not always indicative of whether or not the individual is going to get the disease. Other factors, both genetic and environmental, come into play. The HH heterozygous individual may show disease symptoms that are normally only evident in individuals who have two copies of the mutated gene (homozygous mutant). When this happens, the individual is known as a manifesting heterozygote. Another type is termed compound heterozygote; this is a situation in which the individual has inherited two mutated, but different, forms of the gene [☞ "Allele;" "Homozygote;" "Wild type"].

Histone

Histones are proteins; they exist in five main classes (H1, H2A, H2B, H3, and H4).

Histones H2A, H2B, H3, and H4 come together inside the cell in a specific way, forming what looks like a bead. Base pairs of DNA—in this case 146 base pairs—wrap around this "bead" and then the DNA extends, string-like, over a stretch of 200 bases to the next "bead." The 200-base-pair "string" of DNA is associated with the other histone protein, H1, which helps bind it to the "bead." So DNA is organized in the cell in this "beads on a string"-like appearance.

Another term for the bead in the "bead on a string" structure is nucleosome (H2A, H2B, H3, H4, and 146 base pairs of DNA). Using this "beads on a string" approach, huge amounts of DNA can be packaged into much smaller spaces inside the cell.

Home-brew assay [☞ "Analyte-specific reagent;" "FDA-approved/cleared;" "Research use only"]

Nothing done in the molecular diagnostics laboratory is done in our homes, nor is it vaguely reminiscent of wine- or beer-making, so the term "home-brew" is somewhat unfortunate. Still, it is very descriptive and many in the field use it and instantly recognize the term. A home-brew assay, more professionally known as laboratory-developed or in-house developed assays, are those that have been designed and validated by the clinical laboratory using them for patient care.

These assays are not manufactured, FDA-approved, or FDA-cleared. Their use and development are, however, regulated by the Centers for Medicare & Medicaid Services (CMS) under CLIA '88. The FDA does not regulate home-brew assays, but does require them to be developed using only analyte-specific reagents (ASRs), which are regulated by the FDA. This provides a level of quality and consistency to the assays.

Home-brew assays are commonly used in appropriately certified and credentialed labs and have been extremely beneficial in the field of molecular diagnostics.

Homozygote [☞ "Heterozygote;" "Wild type;" "Allele"] (homozygous)

A homozygote (for a given gene) is an individual with two identical copies of the same gene (alleles) on the two paired chromosomes. This is a result of inheriting identical copies from each parent.

The terms homozygous wild type and homozygous mutant refer to two gene copies encoding the wild-type and mutant proteins, respectively. In some cases, the gene product is colloquially associated with what's called a trait, e.g., eye color, freckles, widow's peak, tongue rolling. In the case of traits, the individual would be considered homozygous, and a sub-designation of wild type or mutant would not apply.

Human Genome Project (HGP)

The Human Genome Project was an internationally collaborative effort to complete the sequencing of all three billion base pairs found in *Homo sapiens*.

Though it took years and billions of dollars to accomplish, the sequence was completed ahead of schedule and under budget. Now that we have the sequence, projects like HapMap [☞ "HapMap"] will facilitate research into human disease that will undoubtedly lead to improved understanding about human biology, refined diagnostics for disease, and more rational and specific therapies, perhaps even the slowing of the aging process. Can we reverse or slow these processes? Disease? Probably. Aging? Maybe. Death? Probably not. Taxes? Absolutely not!

Read about the Human Genome in great detail online (http://www.ornl.gov/sci/techresources/Human_Genome/home.shtml and www.ncbi.nlm.nih.gov/genome/guide/human). The Director of the National Human Genome Research Institute (http://www.genome.gov/) is Dr. Francis Collins, who led the government-based project. The project was completed more quickly than anticipated due to the pressure brought to bear by the private sector, principally a company called Celera, led by J. Craig Venter (who now runs the J. Craig Venter Institutes [see http://www.jcvi.org/]).

Drs. Collins and Venter were the 1998 and 2002 winners of the Association for Molecular Pathology (www.amp.org) Award for Excellence in Molecular Diagnostics, respectively. You can order a free copy of a poster commemorating the HGP (go to http://public.ornl.gov/hgmis/external/poster_request.cfm).

Hybrid

The simple definition of a hybrid is a combination of two or more "things." Living in the Detroit area, a.k.a., the Motor City, we couldn't very well talk about hybrids without mentioning the obvious, hybrid cars, which combine a traditional gasoline engine and an electric motor to produce a vehicle friendlier to the environment and more energy efficient.

In biology, hybrids are created to produce unique characteristics that would not occur naturally. One of the first "created" hybrids was the mule. In fact, the name mule is derived from the Latin word *mulus*, which means the offspring of two different species. The mule is the offspring of a male donkey and a female horse and in some circumstances, such as pulling heavy weights, is preferred over either parent species; mules are sterile.

In molecular biology, a hybrid is the combination of two or more nucleic acid molecules, which would not normally be found in nature. A common hybrid is a double-stranded molecule where one strand is DNA and one strand is RNA (DNA:RNA). One molecular diagnostics technique relies on the creation of DNA:RNA hybrids. Using this technique, DNA from the patient binds to the complementary sequence found in an RNA probe. It is then "captured" onto a solid surface and identified using antibodies that recognize the unusual hybrid entity. This technique is appropriately called Hybrid Capture and is the basis for an FDA-approved technique for detecting the human papillomavirus genome (HPV); HPV is the causative agent of cervical cancer, for which a vaccine now exists.

Hybridization

Hybridization is the process of forming a double-stranded DNA or RNA molecule between a probe (created synthetically in the laboratory) and a target (for example, patient DNA in the clinical molecular diagnostics laboratory).

DNA is naturally double stranded and can be denatured—in other words, made single stranded. If the two strands find each other again, they are said to have reassociated with each other. If, however, the investigator adds a large excess of DNA, called a probe, that has complementarity [☞ "Complementary strands of DNA"] to a particular sequence of interest, the probe, based on its presence in large excess, out-competes the sister strand and finds the target before the sister strand does. This process is called hybridization because a hybrid duplex (target to probe) has been formed instead of simple re-association of the two sister strands [☞ "Denature;" "Duplex;" "Probe"].

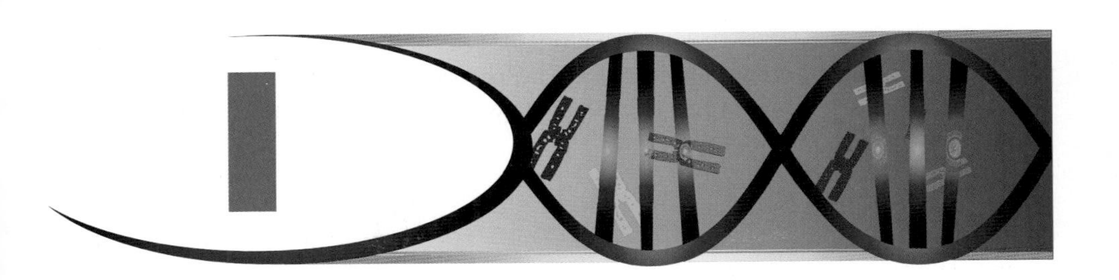

Inheritance

Inherited disorders or traits are passed from one generation to the next through the germline cells. Gregor Mendel [☞ "Mendel, Gregor"] unraveled inheritance patterns in classic breeding experiments using pea plants. These patterns are the basis for our knowledge of how diseases and traits are inherited.

Germline cells, or gametes (sperm, ovum), contain one set of chromosomes or one-half the number of chromosomes present in the rest of the cells in the body (23 chromosomes = one set) [☞ "Germline;" "Somatic cells"]. When a sperm and ovum fuse to create a zygote, the newly formed cell inherits two sets of chromosomes: one set from the mother, one set from the father. Mutations present in one of the inherited chromosomes are passed on; they are inherited.

One might ask, "Why are some members of a family affected by mutations while others are not?" The answer is twofold: first, all of the parent cells contain two full sets of chromosomes: one set from *their* mother and one from *their* father. When gametes are produced, only one chromosome of the pair is transferred (recall that gametes are haploid and contain only 23 chromosomes). If the parent cell has one normal and one mutated chromosome, then the gamete has a 50% chance of getting either the normal chromosome or the mutated chromosome. If the normal chromosome is transferred, then the offspring does not inherit the disease or trait; however, if the mutated chromosome is transferred, the offspring inherits the *potential* to have the disease or trait. This is the second part of the answer, which has to do with whether or not the disease or trait is dominant or recessive in nature.

A dominant mutation refers to the ability of the mutated chromosome to cause a disease or trait all by itself, regardless of whether or not the other chromosome is normal. So, a dominant mutation will cause the disease or trait if present on only one of the inherited chromosomes. If however, a recessive mutation is inherited, then two chromosomes, each with the recessive form of the gene, are required in order for the disease or trait to manifest, in other words, the same mutation has to be inherited from both the mother and the father in order to cause the characteristic in the offspring. The disease or trait would not be present if only one chromosome containing a recessive mutation is inherited.

Examples of a dominant inherited disease and trait are polycystic kidney disease and cleft chin (chin dimple), respectively. Cystic fibrosis and hitchhiker's thumb are examples of a recessive disease and trait, respectively. These examples follow traditional Mendelian inheritance patterns; however, as with everything, there are exceptions. Incomplete dominance is a "blending" of two traits, such as seen in hair texture. In this example, wavy hair is a result of one parent carrying the gene for curly hair and one parent carrying the gene for straight hair with the offspring having a mixture, or wavy hair.

Using this knowledge, molecular diagnosticians can determine whether or not offspring have inherited normal or abnormal genes from the parent(s). This is especially useful for prenatal and carrier screening, or pre-implantation genetic diagnosis, where there is a family history of disease [☞ "Gene tests"].

Initiation codon

[☞ AUG]

In silico

Experimentation in the burgeoning field of pharmacogenomics and indeed genomics itself may sometimes be done by computer analysis of interesting DNA sequences. The "experiments" are said to be done in silico, to reflect the computer-heavy (silicon chips) component. The experiments may involve in silico modeling of how specific gene sequences lead to various gene products and how potential drugs to target them may interact at a molecular, three-dimensional level [☞ "Genomics;" "Pharmacogenomics"].

In situ

An experiment or examination of something in its original position or where it would naturally occur. An example is the examination of a tissue or organ that has been removed from the body but is still intact, such as the identification of fungal DNA sequences within lung tissue.

In utero

In the uterus; used to describe things associated with an unborn child. For example, by examining cells from an amniocentesis examination, one can make statements about genetic diseases to which a fetus may be predisposed while that fetus is still in utero.

In vitro

Experiments or work done in a test tube or some other kind of laboratory container are said to be done in vitro. Mutations may be artificially induced in previously purified DNA, or DNA may be amplified to copious amounts for diagnostic purposes in vitro.

In vivo

Experimental or therapeutic work done inside the body is said to be done in vivo. DNA mutations may be induced in experimental laboratory animals in vivo.

Insertion

In molecular biology, an insertion refers to the addition of one or more nucleotides within a DNA sequence. These additions can be within a gene's coding or non-coding regions and can occur naturally, or during experimental manipulation. Small insertions of one or two base pairs within the coding region can shift the reading frame, resulting in an abnormal protein. These types of insertions are known as frameshift mutations [☞ "Mutation"]. Small insertions outside the coding region can also have dramatic effects, such as on regulation of RNA transcription.

Large insertions may involve the insertion of parts of other chromosomes or whole genes from external sources. Chromosomal insertions (sometimes called chromosomal translocations) occur during cellular division. These types of insertions are commonly seen in cancers such as leukemias and lymphomas [☞ "Chromosomal translocation"]. Insertion of genomic sequences is also seen in some viral infections where the virus inserts all or parts of its genome into the host chromosome. Examples of viruses that do this are Epstein-Barr virus, which causes infectious mononucleosis, Burkitt's lymphoma, and HIV [☞ "Retroviruses;" "Virus"].

Researchers find it useful to insert nucleotides or whole genes into DNA sequences for experimental purposes in a process known as cloning [☞ "Clone; Cloned"]. Small sequences are often inserted into the non-coding regions of individual genes to study the effect on gene expression. Large sequences may be inserted for medical purposes, such as in gene therapy, where it is hoped the inserted sequence will compensate for the defective or missing gene.

Intergenic region

Intergenic regions are the stretches of DNA located between the regions that encode genes, similar to stretches of highways between cities (in this example, city = gene).

Intergenic regions differ from introns in that introns separate sections within a gene; if we continue with the analogy, think of introns as similar to sidewalks separating buildings, where the buildings are analogous to exons [☞ "Intron;" "Exon"].

Some scientists used to call intergenic regions "junk DNA" because they do not contain genes or have an apparent function. While some sequences have been shown to control surrounding gene expression, we do not yet understand the relevance of most intergenic regions. We now believe, however, the designation as junk DNA is not accurate nomenclature [☞ "Junk DNA"].

Intron

Loosely, the opposite of "exon;" introns are intervening stretches of DNA that separate exons, are spliced out of the gene at the RNA level, and are not ultimately expressed as part of the gene product (protein).

For example, consider the following hypothetical gene:

GOBWLTPP?SLPENNCCNCPACJAIL

which gives rise to the following hypothetical protein, MNPLY, that is encoded at the RNA level by BWPPPENNNCPAC.

The KEY is as follows:

GO = "GO" square on the Monopoly® Game Board, analogous to AUG initiation codon [☞ AUG]

BW = Boardwalk
LT = Luxury Tax
PP = Park Place
? = Chance
SL = Short Line Railroad
PENN = Pennsylvania Avenue
CC = Community Chest
NC = North Carolina Avenue
PAC = Pacific Avenue
JAIL = Monopoly Game Board Jail square, analogous to any of three translation termination codons [☞ "Genetic code;" "4-3-20"]

This is the expensive side of the Monopoly game board and in this example, it represents a sequence of hypothetical bases found in every cell of the body that has DNA (basically all except red blood cells). Let's further assume this is the gene that codes for the MNPLY protein that is expressed only in adolescent male brain cells during family vacations. During the right time this gene has to be expressed. To oversimplify, the DNA is transcribed by the cell into RNA. But the RNA that is ultimately translated into the MNPLY protein is much shorter than the gene. That's because the introns (LT, or Luxury Tax; ?, or Chance; SL, or Short Line Railroad; and CC, or Community Chest, all underlined in the sequence above) are spliced out by the cell and the two exons in this gene (the blue and green monopolies of Boardwalk, Park Place, Pennsylvania, North Carolina, and Pacific Avenues) are brought together at the RNA level so that the functional protein, MNPLY is translated [☞ "Exon;" "Junk DNA" (definition 2)].

IUPAC

The International Union of Pure and Applied Chemistry (IUPAC) is an international organization, perhaps best known for defining symbols and abbreviations for chemistries to maintain consistency throughout the world.

Two areas where IUPAC has had an impact on molecular biology are in defining the one- and three-letter codes for all 20 amino acids and the one-letter code and ambiguity codes for the nucleic acids. An example of an IUPAC amino acid designation is for the amino acid alanine, where A is the one letter code and Ala is the correct three-letter designation.

The nucleotides also have the familiar one letter codes (A = adenine; C = cytosine; T = thymine; G = guanine; and U = uracil) as well as codes to designate more than one base. These codes are commonly referred to as ambiguity or degenerate base codes.

Let's say that we have analyzed the same DNA sequence from ten different people. After comparing the sequences, we notice that 50% of the sequences had an A in the 4th position and 50% of the sequences had a C in that position. Instead of writing both an A and a C, we would look up the IUPAC code for this combination and use that code to designate that either of two nucleotides may be present at that position. In this case the IUPAC code is an M for aMino (both A and C contain an amino group) and so anyone viewing the sequence will be able to look up the bases for which M codes, and know that either an A or a C can be found at that position. More information on IUPAC codes is available at http://www.iupac.org/index_to.html [☞ "4-3-20" in the section on numbers].

IVDMIA

In vitro Diagnostic Multivariate Index Assays; an FDA invention for clinical laboratory tests that not only have a technical component (as all tests do) but also have a specialized, non-obvious, software-applied mathematical algorithm through which the test data are "filtered" to generate the final result used for patient management.

For example, many genes may be assessed on an array (DNA chip) and various mathematical manipulations may be applied to the data before a meaningful test result is generated. The first example of a clinical laboratory test cleared under FDA's IVDMIA process was the Mammaprint® test from Agendia in 2007, useful in predicting the likelihood of tumor recurrence in breast cancer patients.

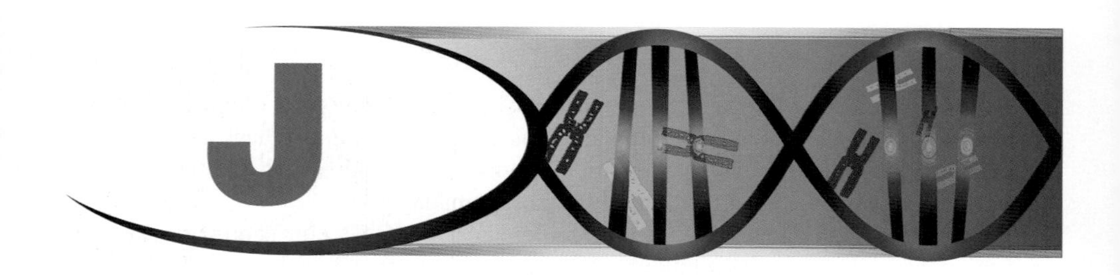

Jumping genes

Formally called transposons; transposons developed the nickname "jumping genes" from their ability to replicate and insert their sequences randomly into the host DNA.

We now know that many transposons do randomly "jump" around, but others have specific spots for insertion into the host chromosome. Transposons have the potential to wreak havoc once inside a cell: they can damage a functional gene if they insert into a coding region; they can damage a gene when "jumping" out by taking extra nucleotides; and they can alter gene expression if they insert into the regulatory region of the gene.

While the genomes of some species are composed of up to 50% transposable elements with no apparent harm, transposons are often deleterious. Diseases such as hemophilia A and B; type II diabetes; and breast, lung, and colon cancers are thought to have transposon-based origins. HIV may have evolved from retrotransposons, and transposons that "infect" bacteria can carry antibiotic-resistance genes, resulting in the numerous multi-drug resistant bacteria we see today.

Junk DNA

Some refer to introns and intergenic regions as junk DNA. This is likely an oversimplification, as introns serve some sort of regulatory or punctuation role within genes, and intergenic regions may control the expression of surrounding genes and are not strictly junk. But then even scientists need to label things they don't quite understand so they can put them in a convenient little compartment in their brains reserved for stuff they don't want to think about that much. Kidding aside, some, perhaps even much, of what we call junk DNA, may have function ascribed to it before long, now that the human genome is fully sequenced and we learn more about it through projects such as HapMap [☞ "Exon;" "HapMap;" "Intron;" "Intergenic region"].

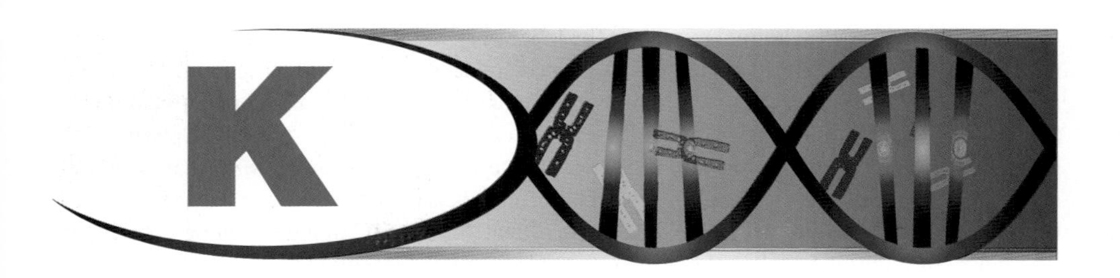

kb

kilobase; 1000 bases. A gene that is 4000 base pairs long is said to be 4 kb (or kbp for kilobase pairs; remember DNA is double stranded) in length.

kDNA

We're not referring to the educational Spanish language public radio station in Washington. kDNA is DNA specific to a particular organelle [☞ "Organelle"], called the kinetoplast. Kinetoplasts are found in protozoa (one-celled pathogens distinct from bacteria) like *Trypanosoma cruzi*, which causes a potentially fatal disease known as Chagas disease, also known as American trypanosomiasis. Kinetoplasts in the organism contain DNA referred to as kDNA; kDNA is highly interconnected or catenated and, at least in part, resembles mitochondrial DNA in that it contains genes for proteins involved in energy production. An electron micrograph of kDNA may be viewed online (see http://www.ebi.ac.uk/parasites/kDNA/Darkfld2.jpg).

Kinase

Kinases are a class of enzymes that add phosphorus molecules to their substrates, thereby turning them "on" or "off" (a substrate is the target molecule with which the business end of the enzyme interacts to speed along a biochemical reaction).

Phosphorus can be made radioactive, and can be obtained commercially and used to "label" DNA for use in the DNA diagnostics laboratory [☞ "Autoradiograph" and "DNA labeling"]. Phosphorylation of proteins is such an important regulatory element and can play such an important role in pathogenesis of disease that all the kinases (and there are many) normally expressed by our cells have been cataloged into something called the human kinome (analogous to human genome). To download a poster, visit http://www.sciencemag.org/cgi/data/298/5600/1912/DC2/1. Many hundreds of genes code for protein kinases.

Knockout mouse

We love dogs. We certainly believe humans should be kind to animals and treat those we use for research and animal husbandry humanely. We do not believe this attitude is inconsistent with the reality that animals may be used as excellent models for human disease in an effort to help eradicate disease and minimize human suffering.

One of the very best animal models is the laboratory mouse, *Mus musculus*, because of its relatively short gestation period and because mice are easy to raise and to keep. Furthermore, as mammals they share much in common with humans at the genomic and biochemical levels. So what we learn in mice may provide lessons we can apply to human biology. Experimental genetics in mice can yield answers in weeks. The entire genome of *M. musculus* has been sequenced and we have the capability to ablate (or knock out) the function of one gene at a time in a very precise manner to see the effects of that "knockout" of a gene in the experimental mouse. In point of fact, knockout mice experimentation preceded the 2002 announcement of the completion of the mouse genome sequence by some 20 years. Similarly, we can add genes in so-called "knock-in" mice to ask related questions. For example, researches have bred a knockout mouse that lacks the gene that encodes a working ion pump in cells lining the lung and gut. This knockout phenotype mimics the human disease cystic fibrosis (CF), and it allows important scientific questions to be asked that lead to better understanding of CF biology and possible improved therapies and cure.

Using these special mouse models aids accumulation of knowledge about a specific gene's role in the body. Severe combined immunodeficiency disease ("Bubble Boy disease"), also known as SCID, is a serious human disease, and is being studied in "knock-in" mouse models. Genes that encode specific elements of the human immune system have been individually inserted into SCID mice to study the result. Such genetic manipulation allows individual components of the complex human immune system to be studied one by one. Similar work is being done to learn more about acute myelogenous leukemia, Huntington's disease, Alzheimer's disease, and many more diseases of humans.

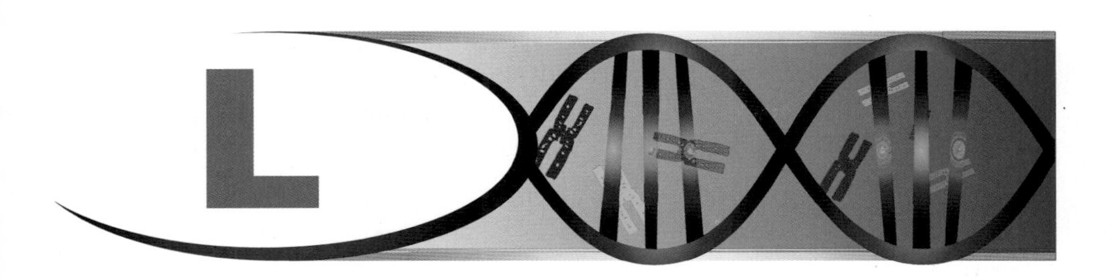

Labeling

[☞ "Autoradiograph;" "Chemiluminescence;" "DNA labeling;" "Kinase"]

Laboratory developed tests (LDT)

[☞ "Home-brew assay;" "FDA-approved/cleared;" "Research use only"]

Lagging strand

The lagging strand refers to a specific component of DNA replication. Growth is dependent on cell division (one cell becomes two, two become four, etc.) and cell division is dependent on DNA replication.

DNA is a double helix, composed of two strands. Think of the two strands as sister strands. When the cell reaches that portion of the cell cycle where it's time to divide, the two sister strands separate slightly in the middle, forming a so-called replication bubble. The molecules and enzymes involved in the business of DNA replication make more DNA bidirectionally. This means that a new daughter strand is made from one sister strand in one direction, and a second daughter strand is made from the other sister strand in the other direction. The daughter strand synthesized in the right-to-left direction is the leading strand. The daughter strand synthesized in the left-to-right direction is the lagging strand. (This is an oversimplification since there is no real right or left in the three-dimensional environment of the nucleus. In this case "right" and "left" refer to a linear representation of DNA one would see on paper, as shown at the following URL: http://medicaldictionary.thefreedictionary.com/_/viewer.aspx?path=dorland&name=fork_replication.jpg).

Lambda (λ) DNA [☞ "Bacteriophage"]

Lambda (λ) phage is a virus that infects bacteria. This virus has a relatively small genome of close to 50,000 base pairs (50 kilobase pairs or kbp) in length. Its genome can be cut into fragments of known size by different restriction endonucleases.

These characteristics make λ DNA a useful tool in the molecular pathology laboratory. When restriction endonuclease-digested λ DNA fragments are electrophoresed [☞ "Electrophoresis"], they migrate to a particular position in the gel, based on their molecular weights. Because we know the sizes of those fragments, we can use them as molecular weight "standards" against which we can "size" DNA fragments whose molecular weights are unknown. This has been common practice in the clinical molecular pathology laboratory for purposes of quality control of the laboratory test results generated. Increasingly, the field is moving to microfluidic assessment of DNA fragment sizes on specialized DNA chips (see http://www.chem.agilent.com/scripts/pds.asp?lPage=1283).

Leading strand

[☞ "Lagging strand"]

Library

A library of books is a collection of books. A library of DNA molecules is a collection of DNA molecules.

In the kind of library we are most used to thinking about, the books are housed in a building called the library. In a DNA library, the DNA molecules are "housed" or contained in vectors. Vectors are carriers of DNA. Commonly used vectors include viruses, yeast artificial chromosomes [☞ "YACs"], and bacterial plasmids [☞ "Plasmid"]. With respect to molecular biology, libraries are created so that one can study, for example, the DNA of a fruit fly (a fruit fly genomic library), or all the transcripts (RNA biochemically manipulated into cDNA) from a mouse liver (a mouse liver cDNA library), or human breast cancer (human breast cancer cDNA library).

LiPA

Line probe assay (see Figure 21).

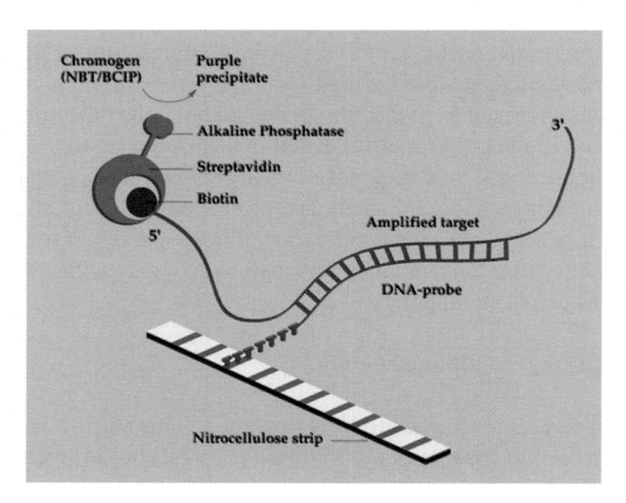

Figure 21. The solid support with DNA probes bound along its length; in this example an amplified target has hybridized to one (exaggerated for clarity) of the bound probes. Because the amplicon is "biotinylated" (tagged with a reporter molecule), chemical reactions depicted on the top left of the figure generate a purple color at the site of hybridization, indicating a positive result for the known probe bound at that site. (In this example, only the third probe from the top is positive; the other locations on the strip appear purple to illustrate the line probe concept, but would not signal positive-or purple-in the assay.

Reprinted with permission from Innogenetics, Belgium (http://www.innogenetics.be/default.asp).

In traditional hybridization assays like the Southern blot, the target DNA is treated in such a way that it is available for detection by a probe [☞ "Southern blot"]. The idea behind the line probe assay (and the reverse dot blot) may be thought of as the opposite, or reverse, of that strategy. Unlike the Southern blot, manufacturers of line probe assays and reverse dot blots bind the specific DNA probes of interest to the solid support (often a tough piece of nylon strip). These probes may be many species of a particular bacterial genus (for example, *Mycobacterium tuberculosis*, *M. kansasii*, *M. leprae*, etc.) or dozens of different disease-causing mutations within a gene, e.g., the *CFTR* gene which, when mutated, causes cystic fibrosis. To those strips is added purified (from a patient specimen) DNA that has been amplified, generally using PCR [☞ "PCR"], for the potential target(s) of interest. If, using the bacterial example above, the specimen was positive for *M. tuberculosis* and not any of the other species of *Mycobacteria*, the specimen DNA would only bind to the fixed *M. tuberculosis* probe. Through a series of reactions, that particular hybridization event is observed at the site of hybridization on the nylon strip. The probes are arranged in a series of lines; in this example, one line is observed. By lining that up with a key, one can deduce which pathogen has infected the specimen. Reverse dot blots are similar in nature and the probes are "dotted" onto the membranes, generating a circular, dot-shaped reaction.

These are methods to search for a "needle in a haystack." While PCR makes a haystack full of needles, those needles still need to be observed, and so the lines and dots on the LiPAs and dot blots, respectively, may be thought of as permanently affixed magnets that pull the needles out of the haystack where they then stick and can be observed.

Locked nucleic acid (LNA)

TECHIE: Normally, nucleic acids are flexible, a characteristic that allows the molecule to easily undergo cellular processes such as replication and transcription. A locked nucleic acid (LNA) is an RNA nucleotide that has been modified through the addition of a bridge, which binds the 2'-O to the 4'-C, locking the nucleotide into a non-flexible conformation. Because of this modification, when LNA nucleotides are incorporated into oligonucleotide probes, they exhibit a higher affinity for the target nucleic acid, either DNA or RNA, increasing sensitivity and specificity [☞ "Oligonucleotide;" "Probe"].

Oligonucleotide probes containing one or more LNA bases can be used in lieu of normal probes. One benefit of LNA probes is that they can be shorter, as short as 13 base pairs. Shorter probes may be able to detect targets where long probes cannot, due to limited consensus sequences [☞ "Consensus sequences"]. Other benefits of LNA-containing probes are that they are better at binding difficult AT- or GC-rich regions and they have greater thermal duplex stability (increased melting temperature), which improves the reaction specificity.

Locus

Locus means place or position; genetically speaking, the term refers to a gene's position on a chromosome. So the locus for a particular gene in the human genome is the place where you would find that gene.

Jargon has also turned the word “locus” into a synonym for group, as in the HLA locus of genes (the HLA locus is involved in human immune response and an important criterion for “matching” potential organ donors and recipients) [☞ “LOH”].

LOH

Loss of heterozygosity.

Genes have two alleles. If the two are identical, then an individual is said to be homozygous at that particular locus. If the homozygous alleles are wild type or normal, then the individual is homozygous wild type. If the two identical alleles are both mutated, then the individual is homozygous mutant. If the individual has one of each allele, s/he is heterozygous for that mutation.

Imagine a particular genetic locus heterozygous for a harmful mutant allele and a dominant normal allele; a mutation in the normal allele leaves only the deleterious mutation. Disease would then result, due to so-called loss of heterozygosity. Heterozygosity was holding disease in check and when it’s lost, disease ensues.

Examples of diseases associated with LOH include some forms of colorectal cancer and retinoblastoma, a tumor of the eye.

Luminol

Luminol has become commonly known to viewers of TV shows such as CSI (Crime Scene Investigation), NCIS (Naval Criminal Investigative Service), and Forensic Files. In these shows, the crime scene investigator sprays a solution containing luminol on a suspected blood stain, illuminates the area with UV light, observes a blue glow, and before the last commercial break, the perpetrator has been identified and caught.

In reality, luminol is successfully used in forensic analysis—however, not as freely as on TV. First of all, luminol reacts with compounds other than blood. For blood, the reaction that causes the blue glow is due to the compound hemoglobin, which is found in red blood cells. Hemoglobin acts as a catalyst between the luminol reagent and other chemicals present in the solution, causing the light-producing chemical reaction formally known as chemiluminescence (the blue glow). Other compounds such as copper, iron peroxides, cyanides, horseradish, some vegetation, some bleaches, and cleaning supplies also act as catalysts, so whether the blue glow is specifically being caused by the presence of human blood (animal blood will also react) needs to be determined. Second, luminol may destroy some evidence and should only be used after all other forensic analysis has been completed. This is where molecular biology comes into the story. Luminol may be used to detect the presence of minute amounts of blood, such as blood spatter, which may not be detectable by other means. Since luminol does not affect subsequent DNA analysis, this allows the crime scene investigator to find and recover the evidence, perform DNA analysis, and use that to further the investigation.

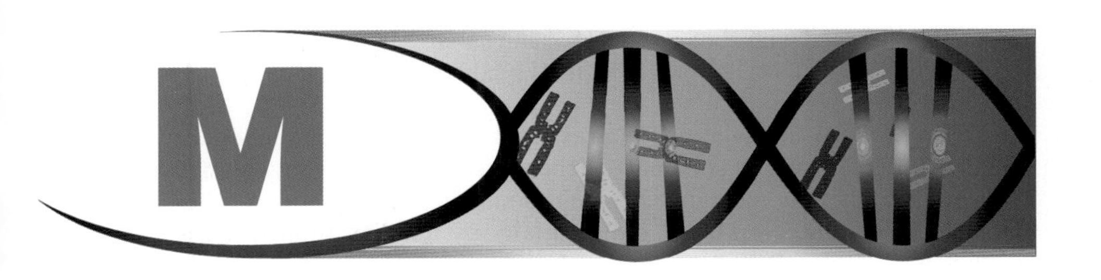

Major groove [☞ "Minor groove"]

When things are going just great, you're in a major groove. With respect to DNA however, the major groove is the larger of the two indentations repeated in a regular fashion throughout the DNA double helix (see Figure 8 on page 16).

Melting temperature

[☞ "T_m"]

Mendel, Gregor (1822–1884)

An Augustinian monk who worked out many of the concepts of heredity and heritable traits using pea plants that he pollinated and bred at the monastery.

In 1865, Mendel published his work, "Experiments on Hybrid Plants," in the *Proceedings of the Natural History Society of Brno*. Mendel's is considered classic work. To honor Mendel, classic genetics is sometimes referred to as Mendelian genetics.

We now know about more complicated inheritance patterns found in nature, and these are "variations on the theme" of Mendelian genetics. Most often, however, genetic characteristics, including many diseases, follow the rules demonstrated by Mendel in pea plants, and these traits are said to be inherited in classic Mendelian fashion [☞ "Inheritance"].

Methylation

To methylate something is to add a methyl group ($-CH_3$).

In the case of DNA, the methyl group is added to a cytosine residue when it is next to a guanine residue. When these two residues are next to each other, they are known as CpG dinucleotides (where p = the phosphodiester bond). Methylation of these CpG residues is almost always found in the DNA sequence before the gene, known as the promoter region, which is responsible for gene transcription (how much of the gene is made or expressed). When methylated residues are present, gene transcription is almost always reduced; thus, methylation is a normal mechanism of gene regulation.

DNA methylation plays an important role in cancers and diseases by altering expression of genes that regulate cell growth. In other words, the proteins that keep cells from growing uncontrollably and becoming a tumor are no longer made, allowing cells to proliferate in a cancerous fashion. Cancers such as stomach, lung, prostrate, melanoma, and colon have been associated with abnormal DNA methylation patterns. In addition, diseases such as Prader-Willi, Angelman, Rett, and Fragile X syndrome result from abnormal DNA methylation patterns or processes.

One way DNA methylation patterns can be detected in the laboratory is to treat single-stranded DNA with sodium bisulfite. This chemical destroys all non-methylated cytosines, so that only methylated cytosines remain. Identification of these remaining cytosines can be accomplished preferentially by amplification using a specialized "methylation-specific PCR," or through identification using other methods such as direct DNA sequencing or single-nucleotide primer extension assays.

micro RNA (miRNA)

[☞ "RNA"]

Microarray

[☞ "DNA chips"]

Microsatellite DNA [☞ "CODIS;" "Intron;" "Junk DNA"]

Just as satellites orbit the Earth, microsatellite DNA (sometimes called simply satellite DNA) has been so-named because it "orbits" or surrounds coding DNA. (We put "orbits" in quotes because the word implies that satellite DNA is somehow physically separate from DNA and that is not the case.)

Microsatellite DNA is made up of short tandem repeats (STRs), which are repetitive stretches of DNA base pairs two to five nucleotides in length. For example, the STR made up of the nucleotides C, A, and T may repeat many dozens of times in tandem—

CATCATCATCATCAT

—*ad nauseum.*

These STRs appear scattered throughout the human genome and are found in "junk" DNA and introns "orbiting" the coding regions or exons, hence the name. These regions have use in identity testing and in disease analysis because they tend to be naturally unstable and therefore prone to mutation. This inherent instability is referred to as microsatellite instability [☞ "Microsatellite instability"].

Microsatellite instability (MSI)

MSI is a disagreement in size of a particular microsatellite region between, for example, tumor and normal tissue from the same patient.

MSI is thus a marker of a biochemical defect manifested in the tumor where errors in DNA replication that would otherwise be proofread and corrected by the cell go uncorrected. MSI may be used diagnostically, and because the sequences of the satellites in question are specific to the tumor, MSI may be used as something of a genetic signature to monitor the tumor during therapy for response and possible recurrence.

There is also evidence to suggest that patients whose tumors express MSI may be candidates for different regimens of chemotherapy than those whose tumors do not express MSI. This is an example of the notion of personalized, tailored therapy based on molecular characteristics.

Miescher, Friedrich (1844–1895)

This often-overlooked Swiss physician/scientist discovered and isolated DNA for the first time in 1869.

At the time, Miescher was working on lymphocytes, and the isolation of DNA was unintentional. Miescher also suggested that his isolated substance may somehow be involved in determining traits and characteristics several decades before DNA was formally identified as the hereditary material within a cell. Miescher used as starting material (OK, here's fair warning—you may not want to go on if you've eaten recently or are considering eating soon) pus and salmon sperm (really!) and named the stuff that he isolated from the nuclei of cells nuclein. The subsequent designation deoxyribo*nucleic* acid is derived from this name.

Minor groove [☞ "Major groove"]

When things are going pretty well but not as great as when you're in a major groove. The DNA double helix makes a complete turn and begins a new one every 10 bases. The way the two strands of nucleotides that form the DNA double helix wind around each other causes grooves within the molecule, one relatively small indentation or groove, and a larger one. The small one is the minor groove and the larger one is the major groove (see Figure 8 on page 16). Some also call these the shallow and deep groove, respectively.

Minor groove binding protein

DNA is a double helix with two very distinct grooves; a major groove and a minor groove. [☞ "Major groove;" "Minor groove"] Proteins interact with DNA and can bind specifically to parts of, and/or within, either groove. Proteins that preferentially bind to the minor grove are called minor groove binding proteins. These proteins participate in many cellular processes such as transcription and DNA repair.

TECHIE: In the lab, minor groove binding proteins may also be attached to probes, such as the MGB Eclipse™ TaqMan probes. The presence of the MGB protein increases the stability of the probe, allowing shorter probes to be designed. This is especially beneficial for sequences problematic for binding by probes [☞ "Probe"].

Mismatch

With respect to DNA, mismatch refers to lack of complementarity.

DNA is composed of nucleotides: adenine (A), cytosine (C), thymine (T), and guanine (G). The rules of base pairing dictate that, in DNA, A always base pairs to T and G always base pairs to C; so goes the double helix. Usually, though, when you see the word "always" you know the author's about to tell you about the exceptions. Well, yes, mistakes happen, mutations occur, and a G can find itself across from a T, or a C may find itself mismatched with not a G, but an A. That's mismatching, it's to be avoided, and our bodies have evolved a biochemical (proofreading) way to not only minimize the occurrence of mismatching but also repair it when it occurs [☞ "Complementary strands of DNA;" "Microsatellite instability"].

Mitochondrial DNA (mtDNA)

Mitochondria are distinct elements within animal and human cells (Figure 22). Such a distinct element or subunit of the cell is called an organelle (the nucleus is another example of an organelle). Mitochondria are involved in oxygen transfer and energy conversion.

Many remember first learning about mitochondria in elementary school as the "powerhouse of the cell." Mitochondria contain their own DNA (this is distinct from the DNA in the cell's nucleus). Each cell has hundreds to thousands of mitochondria and each mitochondrion holds up to 100 mitochondrial DNA copies.

Mitochondrial DNA is abbreviated mtDNA. The mitochondrial genome, only 16,569 base pairs in length, is small compared to the nuclear genome. There are no introns in mtDNA [☞ "Intron"]. Human mtDNA has 37 genes. The mutation rate of mtDNA is greater than that of nuclear DNA, and mtDNA is derived only from one's mother. Mutations in mtDNA are known to cause human disease, particularly in the brain, heart, liver, kidney, muscle, and pancreas. Examples include Leber's hereditary optical neuropathy, Pearson syndrome (bone marrow and pancreatic failure), and myoclonic epilepsy with ragged red fibers.

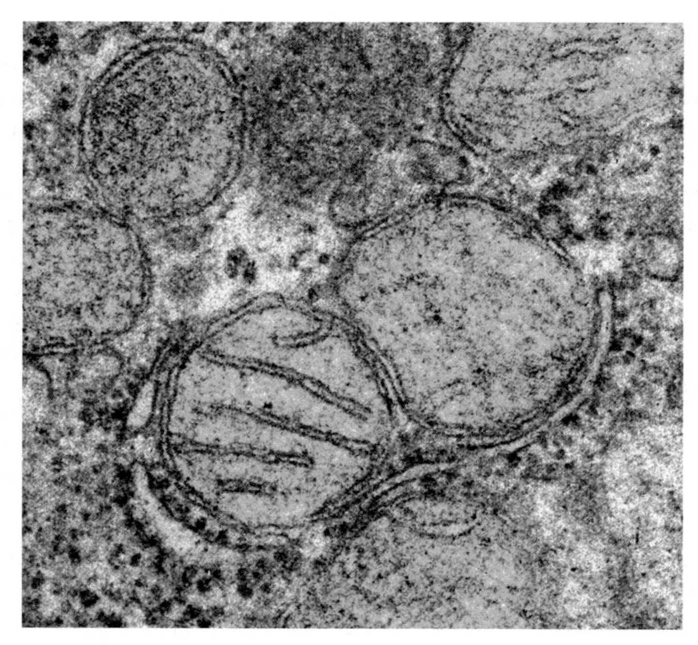

Figure 22. Mitochondria.

Reprinted with permission from Tina Carvalho and the MicroAngela website (http://www.pbrc.hawaii.edu/bemf/microangela). MicroAngela is a creation of Tina (Weatherby) Carvalho of the Biological Electron Microscope Facility, (BEMF), part of the Pacific Biomedical Research Center at the University of Hawaii at Manoa.

Examination of mtDNA is being used more and more in forensic examination; learn more at http://www.ornl.gov/TechResources/Human_Genome/elsi/forensics.html [☞ "Heteroplasmy"].

Molecular beacons

[☞ "Beacons"]

Molecular biology

Molecular biology is the field in which both of this book's authors received doctoral degrees. Molecular biology is the study of the business of life at the level of the lowest common denominator, the molecules that carry it (life) out: DNA, RNA, and proteins.

Molecular biology as applied in the clinical diagnostics laboratory has been given the moniker of "molecular pathology" or "molecular diagnostics." Applications to genetics are called "molecular genetics."

Though proteins are, of course, molecules, molecular pathology has traditionally focused on diagnostics using DNA and RNA. As the study of proteomics (global, systemic protein pattern examination) moves closer to clinical applications, molecular pathology may expand into the protein realm.

Molecular pathology

Pathology is the study of those stimuli that cause disease and the examination of tissue affected by disease. Molecular pathology (aka molecular diagnostics) is the application of the tools of molecular biology (DNA technology) to the medical practice of diagnostic pathology.

The Association for Molecular Pathology (www.amp.org) is the professional society devoted to molecular pathology, a group of which one of the authors (DHF) was proud to serve as President in 2003.

Molecular profiling

[☞ "Expression profiling"]

Monosomy

Normal somatic cells contain 46 chromosomes; 23 pairs. When one member of a pair is missing, this is called monosomy. When only part of one chromosome pair is missing, this is called partial monosomy.

Perhaps the most common monosomy syndrome is Turner syndrome, where one of the X chromosomes in females (males only have one X chromosome—and a Y chromosome) is missing. Turner syndrome affects females and is characterized by a short and stocky stature, webbed neck, congenital heart disease, and the absence of sexually mature organs, among other traits. Turner syndrome does not affect intelligence; however, some learning disabilities are associated with these patients [☞ "Somatic cells;" "Ploidy;" "Trisomy"].

mRNA

Messenger RNA; DNA is transcribed into mRNA, which is ultimately translated into protein.

MRSA

MRSA is an acronym for Methicillin Resistant *Staphylococcus aureus*.

Methicillin is an antibiotic, and *S. aureus* (we commonly drop the entire genus name and abbreviate it with just the first letter followed by the species name, in italics) is a bacterium that causes a wide variety of nasty infections. MRSA has made a lot of news lately because it is a major cause of community-associated (CA-MRSA) disease as well as hospital-associated (HA-MRSA; also called healthcare-associated) infections. (Nosocomial infections are defined as those acquired in the hospital.) CA-MRSA infections are best known for causing severe skin infections such as pimples, boils, and abscesses. The infections are commonly transmitted by skin-to-skin contact and by contact with contaminated items (clothing, towels, bedding); surfaces (changing tables, wrestling mats, sauna benches, exercise equipment, locker rooms, door knobs); and objects (hot tubs).

Because of the mode of transmission, MRSA is commonly found among students, athletes, military personnel, and prison inmates. In January 2007, for example, several Miami Dolphins' football players contracted CA-MRSA infections through cuts in their skins, probably through contact with contaminated Astroturf. Other football teams such as the Washington Redskins, the St. Louis Rams, the Tampa Bay Buccaneers, and the Cleveland Browns as well as professional hockey, baseball, and basketball teams have also had to deal with CA-MRSA infections. CA-MRSA infections can infect healthy individuals of any age group.

Patients at high-risk for HA-MRSA infections include those in intensive care units, those undergoing surgical procedures, the immuno-compromised, dialysis and chemotherapy patients, and patients in long-term care facilities. HA-MRSA infections can be localized (e.g., at the surgical wound site) or systemic (e.g., in the blood stream, cerebral spinal fluid, etc.) and can affect almost all organs and systems including the lungs (pneumonia), heart (endocarditis), bones (osteomyelitis), and central nervous and digestive systems.

Methicillin resistance is due to the addition of a new gene that inserts itself into the *S. aureus* chromosome (bacteria have but a single choromosome). This new gene, *mecA*, encodes a protein that confers resistance to methicillin and other penicillin-based antibiotics. Testing for the presence of this gene in *S. aureus* can be done easily and rapidly using molecular techniques including PCR. Because individuals may harbor (be colonized with) MRSA without any apparent signs or symptoms, many hospitals have implemented MRSA screening programs in an attempt to control transmission of the organism. The screening programs are designed to test all inpatients and outpatients prior to admission or procedures to learn if they carry MRSA. Using one of the rapid molecular methods (see Table 3 under "FDA" entry on page 53) greatly facilitates the screening process. If the patient is found to be a carrier of MRSA, s/he is treated and isolated from the normal patient population, thus preventing spread of the organism to other patients as well as to hospital personnel [☞ "Antibiotic resistance"].

Multiplex Ligation-Dependent Probe Amplification (MLPA)

TECHIE: This may partially be considered another competitor in the "PCR Wannabe" category. Why partially? Because MLPA is actually a combination of the oligonucleotide ligation assay (OLA) and PCR.

Like OLA, MLPA involves the joining of two oligonucleotide probes that bind to the target sequence immediately adjacent to each other. The 3' end of one probe and the 5' end of the adjacent probe must be exact matches to the target sequence in order for the joining (ligation, molecularly speaking) to occur. Different combinations of probes and target sequences may be included to create a multiplex assay.

In MLPA, PCR amplification occurs through the addition of universal primer binding sites to each probe. The first probe has the universal PCR primer site added to the 5' end of the probe (the 3' end binds the target) and the second probe has the PCR primer site added to its 3' end (the 5' end binds the target) [☞ "Primer"]. By attaching the universal primer binding sites to all probe combinations, exponential amplification of each target can occur using only one primer set.

MLPA assays are useful for determining single nucleotide changes as well as gross chromosomal deletions or duplications in clinical and research laboratories [☞ "Oligonucleotide ligation assay (OLA);" "PCR"].

Mutation

Mistake. A mutation in DNA is an error or permanent alteration that has occurred in the coding sequence of a gene or genetic regulatory element.

Such errors occur during DNA replication and may be a result of environmental insult that has somehow disturbed the DNA sequence or altered the cell's ability to correct those errors. Mutation may also be due to an error introduced naturally by the DNA copying machinery of the cell. Sometimes mutations are advantageous and have been introduced "on purpose" by nature in an effort to deal with a particular problem (like too many malaria-causing insects flying around; carriers of the sickle cell anemia [SCA] mutation are somewhat protected against malaria). Mutations help species evolve as necessary and are the biological mechanism behind Charles Darwin's theory of "Survival of the Fittest."

To amplify on the example given above, children with the mutation that causes sickle cell trait have natural resistance to a fatal form of malaria in Africa (sickle cell anemia is caused by two mutations in a particular gene and is a serious clinical condition; those with one normal copy of the gene and one mutated copy of the gene do not have SCA but rather sickle cell trait).

Of course mutations are not necessarily a good thing. Mutations can have a range of effects on organisms, from no effect (silent mutation) to carrier status to deleterious and damaging effects like outright disease or death (some mutations that occur in utero are incompatible with life).

There are several classes of mutations. They are: nonsense, missense, frameshift, insertion, deletion, and point mutations (Table 5).

Table 5. Types of Mutations

Type of Mutation	Description
Nonsense	Erroneously introduced into the reading frame of a gene is a codon for STOP such that the growing protein chain is prematurely terminated; protein is shorter than normal and may be partially functional, largely functional, or totally nonfunctional. β thalassemia, a chronic anemia, is caused by many different kinds of mutations, some of which are nonsense mutations in the gene for β globin.
Missense	The DNA coding sequence for the gene has had one triplet codon altered such that a different amino acid is substituted for the one that is typically present. Some mutations in the low density lipoprotein receptor gene are missense mutations and cause familial hypercholesteremia, which leads to coronary heart disease.
Frameshift	The introduction of some number of bases, not divisible by three, into the reading frame of a gene. Codons in a gene are composed of three bases. So if, for example, one base is added or four are deleted, the order of the amino acids encoded by that gene has been wrecked. The result may be a prematurely truncated protein (if a STOP codon is created where there wasn't one before) or a protein that has little relationship at the amino acid level to the normal protein because it is composed of a very different amino acid sequence. The result is usually a bad one. The 185delAG mutation is an example. In this mutation, two bases, adenine and guanine, are deleted from exon 2 of *BRCA1*, altering the translational reading frame of the subsequent mRNA. The frequency of 185delAG in Ashkenazi Jewish women, for example, is one in 107, and the mutation is associated with the onset of breast cancer in this group before the age of 40. This mutation is a candidate for screening in this population.
Insertion	One or more bases are inserted; a one- or two- (or any number not divisible by 3) base pair insertion is a kind of frameshift mutation.
Deletion	Bases are deleted; a deletion can also cause a frameshift mutation if the number of bases deleted is a number not divisible by three. The ΔF_{508} mutation, one of those that causes cystic fibrosis, is a deletion mutation.
Point	A single base pair has been changed; it can be a substitution, insertion, or deletion. Substitutions that don't change the amino acid encoded by the triplet codon are silent mutations. For example, GCC codes for the amino acid alanine. So does GCA, so if that mutation occurs (C mutates to A at position 3) there's no effect on the gene product (protein). Some point mutations, e.g., the single base change that result in sickle cell anemia are deleterious.
Splice site	☞ "Splicing"

[☞ "Codon;" "Genetic code;" "Genotype;" "ORF"]

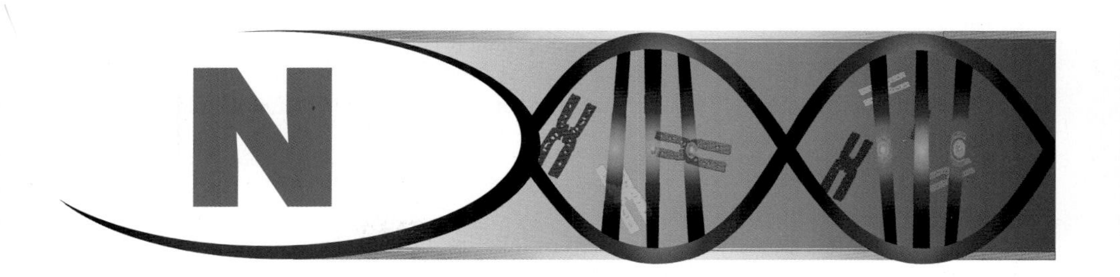

Naked DNA

Naked DNA, as the term applies, is nothing but purified DNA molecules in an appropriate water-based buffer. Naked DNA has been shown to be viable in accomplishing gene therapy. Several mouse models have been used in which naked DNA, injected directly into the diseased tissue, has reduced disease symptoms [☞ "Gene therapy"].

NASBA™ (Nucleic Acid Sequence Based Amplification)

NASBA is an in vitro nucleic acid amplification technique, a PCR "wannabe."

NASBA™ is the biochemistry behind a suite of products marketed by bioMérieux, a French in vitro diagnostics company. Unlike PCR, which accomplishes amplification of input DNA by cycling among different temperatures to accomplish the necessary biochemistry, NASBA is isothermal; all phases of the biochemistry occur at a single temperature (approximately 40 °C).

TECHIE: NASBA, which is particularly well-suited to RNA amplification, accomplishes nucleic acid amplification using three enzymes: (1) reverse transcriptase [RT; ☞ "Reverse transcriptase" for a description of this enzyme's action]; (2) RNase H; and (3) T7 RNA polymerase. Two primers are also used [☞ "Primer"] to initiate the different reactions that occur. NASBA technology may be used to detect the presence of RNA-containing viruses, e.g., hepatitis C virus or HIV. RT forms DNA from viral RNA, if present in the patient specimen, using primer number one. The RNA in the resultant RNA:DNA hybrid is destroyed by RNase H, an enzyme that specifically chews up RNA in such hybrids (hence the "H"). The remaining RNA participates in further reactions using RT and T7 RNA polymerase (a bacterial virus called T7 is the source of this RNA polymerase) and the second primer to generate an exponential amplification of the RNA in about 90 minutes. Typical amplification is on the order of a billion-fold increase in the amount of input RNA. Two of the primary advantages of this technique over PCR [☞ "PCR"] for RNA are that (1) one can use RNA without having to first turn it into cDNA [☞ "cDNA"], and (2) the NASBA reaction can proceed at one temperature, eliminating the need for expensive thermal cycling devices; one can use an ordinary water bath. Real-time NASBA may also be performed with specialized equipment that combines amplification and detection.

NCBI (National Center for Biotechnology Information)

NCBI was established in 1988 as a resource center for "molecular biology information" for scientists and the public alike. It is administered by the National Library of Medicine under the National Institutes of Health.

NCBI maintains teaching and research tools, databases including GenBank, the BLAST search program, PubMed search engine, as well as other resources. For a further listing of offered services visit the NCBI website (www.ncbi.nlm.nih.gov) [☞ "GenBank"].

Nick translation

Nick translation is an enzymatic method of labeling DNA probes with radioactively or otherwise tagged deoxyribonucleotides that are incorporated into newly made DNA molecules during the course of the reaction.

Once tagged, these DNA probes can be used as reporter molecules in subsequent tests to answer questions about DNA or RNA obtained from a patient specimen; the technique is becoming increasingly obsolete, particularly in the clinical laboratory setting (see Figure 23) [☞ "Autoradiograph;" "Chemiluminescence;" "DNA labeling;" "Oligonucleotide priming;" "Southern blot"].

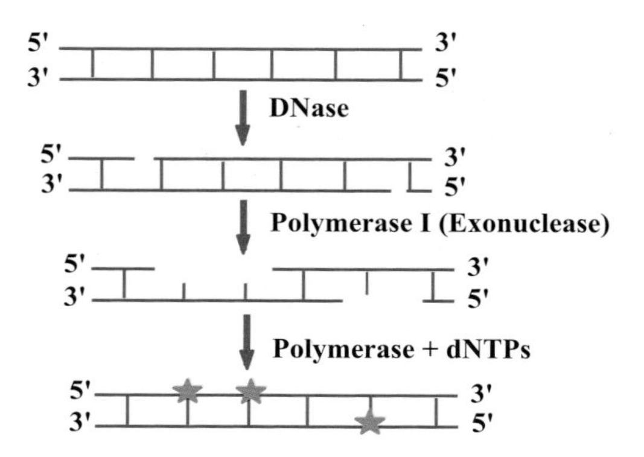

Figure 23. DNA is nicked with DNase, the nick is made larger with exonuclease, and then the polymerase activity of DNA polymerase "repairs" the damage using labeled dNTPs. The stars represent the reporter molecules on the labeled dNTPs.

Reprinted with permission from Tsongalis GT, Coleman WB. Molecular diagnostics: a training and study guide. Washington, DC: AACC Press, 2002.

Northern blot [☞ also "Southern blot"]

The northern blot is essentially the same thing as the Southern blot, so see that entry for a more intensive explanation. The key difference is that while in Southern blotting the target of investigation is patient DNA, in the northern blot the target is RNA.

Unlike Southern blots (which used to be a mainstay of the molecular diagnostics laboratory but are becoming increasingly obsolete), northern blots were never routinely used in the clinical molecular pathology laboratory. Also, while the Southern blot was developed by a person, Dr. Edwin Southern, there was no "Dr. Northern," which is why one capitalizes Southern blot but not northern blot.

Northern blot derives its name from the idea that it is, in a way, the opposite of the Southern blot, with respect to the target of investigation, although RNA is not strictly the "opposite" of DNA.

Nuclease (DNase; RNase)

A nuclease is a protein whose biochemical job is to digest nucleic acids or nucleotides.

A nuclease can be RNA specific (RNase) or DNA specific (DNase). Action can be internal to the nucleic acid (endonuclease; endo for inside or within) or from the end where the nuclease "chews off" one nucleotide at a time (exonuclease; exo for outside or at the end). These enzymes have been co-opted by molecular biologists for use as tools in the molecular biology laboratory. Restriction endonucleases are used all the time. DNase I has a place in nick translation [☞ "Nick translation"]. RNase H degrades the RNA strand in a cDNA:RNA hybrid. There are numerous other examples.

The pancreas is rich in nucleases. Nucleases have a normal role in the body, which is to enzymatically break down any nucleic acids ingested during eating. This is why no one ought to fear ingesting genetically engineered tomatoes, for example. Degradation of nucleic acids occurs in the intestine using the nucleases secreted by the pancreas. Rattlesnake and Russell's viper venom also contain nucleases that are not particular; these enzymes work equally well to degrade DNA or RNA, and of course do not have a natural role in the body.

Nucleases can be a problem when extracting and purifying nucleic acids in the laboratory. DNases, and more commonly RNases, are found in the environment and can contaminate specimens and reagents, destroying the nucleic acid, making it unavailable for further laboratory analysis. Special precautions must be taken to avoid contamination with these nucleases. In addition, DNA purification for molecular pathology investigation of tissues like pancreas or nuclease-rich tumors must proceed quickly to inactivate nucleases before they can act on the DNA and RNA present.

Nucleic acids

Simply put—what this book is about. Nucleic acids are one class of naturally occurring biochemical entities (the other three are proteins, lipids, and carbohydrates). Deoxyribonucleic acid (DNA) and ribonucleic acid (RNA) are the two prime examples.

Nucleic acids are composed of sugar molecules, nitrogenous bases, and phosphate groups. When one of each of these joins, a nucleotide is formed. When nucleotides chemically join to each other, nucleic acids are formed. If the sugar molecule is a ribose-containing sugar, the nucleic acid formed is RNA. If the sugar molecule is a deoxyribose (missing one oxygen molecule)-containing sugar, the nucleic acid formed is DNA.

Nucleoside

[☞ "Nucleotide"]

Nucleosome [☞ "Histone"]

DNA wound around histones forms a structure known as a nucleosome, important in the compression of DNA, so that the very long DNA strands present in a nucleus physically "fit" there.

Nucleotide

In the same way that amino acids are the building blocks of proteins, nucleotides are the building blocks of the nucleic acids, DNA and RNA.

Nucleotides are composed of phosphate groups, a five-sided sugar molecule (ribose sugars in RNA; deoxyribose sugars in DNA), and nitrogen-containing bases. These bases fall into two classes: (1) pyrimidines and (2) purines. Pyrimidines are chemically distinct from purines and include cytosine (C), thymine (T), and uracil (U; a base found only in RNA). Purines include adenine (A) and guanine (G). A nucleotide without its phosphate group is called a "nucleoside."

Nucleotides are often written as abbreviations: AMP, CTP, NTP, etc. The codes for these abbreviations are as follows:

- Letter # 1 stands for the base: A, C, G, T, U or N for aNy of the bases
- Letter # 2 is M, D, or T for mono, di or tri (indicative of the presence of one, two, or three phosphate groups)
- Letter # 3 is always P for phosphate

If this three-letter abbreviation is preceded by a lower case "d," that is a designation for a deoxyribonucleotide (a DNA building block). If there is no "d" prefix, a ribonucleotide (an RNA building block) is understood.

EXAMPLES:

- AMP: adenosine monophosphate
- dTTP: deoxythymidine triphosphate

See Figure 24.

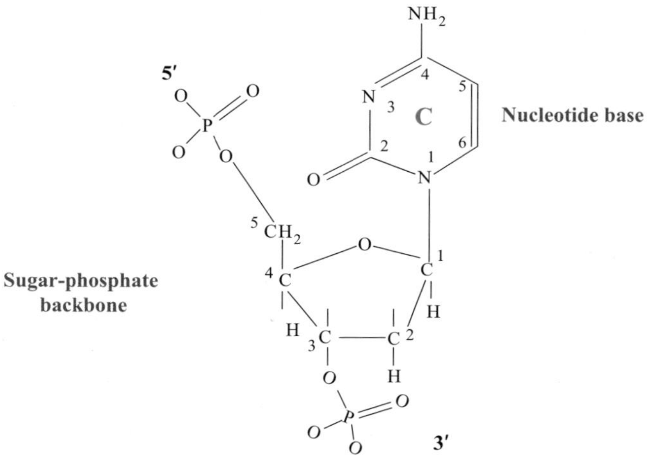

Figure 24. A nucleotide. Note the phosphate group on the top left linked to a five-member sugar ring linked to another phosphate group that links to another (unseen) sugar in the sugar-phosphate backbone of DNA. The sugar is linked to a nitrogenous base, in this case cytosine (C). The combination of the phosphate, sugar, and base is a nucleotide.

Reprinted with permission from Tsongalis GT, Coleman WB. Molecular diagnostics: a training and study guide. Washington, DC: AACC Press, 2002.

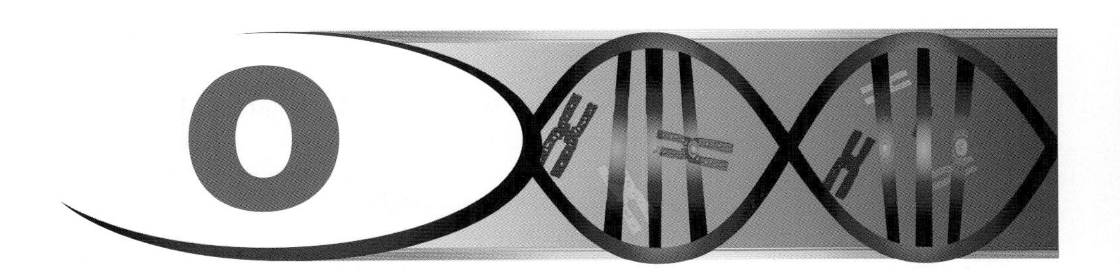

Oligonucleotide

"Oligo" is derived from the Greek for "few." If you string together a few nucleotides, then you have an oligonucleotide or an oligo (slang).

Oligos are absolutely essential components of virtually all molecular diagnostics laboratory clinical assays and molecular biology laboratory experiments. Oligos are used as both primers and probes and can be labeled with chemicals or fluorescent molecules for subsequent detection.

With appropriately sophisticated instruments, oligos can be synthesized using the natural nucleotide bases or alternate bases, and other modifications can be incorporated to increase sensitivity and specificity. Most laboratory scientists, however, do not choose to make their own oligos. Instead they take advantage of the competition in the oligo synthesis marketplace and call someone's toll-free number or access a website and order the specific oligo needed for their experimentation or assay development, and presto—it arrives by courier the next day for about a dollar a base [☞ "PCR;" "Primer;" "Probe"].

Oligonucleotide arrays

[☞ "DNA chips"]

Oligonucleotide ligation assay (OLA)

TECHIE: The oligonucleotide ligation assay (OLA) is a highly sensitive technique that results in a linear (as opposed to exponential, like PCR) amplification of the target sequence.

The OLA technique employs two sequence-specific probes designed to anneal immediately next to each other on the target sequence. If both probes anneal to the target, they are then joined together using the enzyme, ligase, creating a detectable product.

The key to the OLA assay is that the probes must exactly match the target sequence on their adjacent ends in order to be joined together to create the final product. If the two sequences are not exact matches, then they will not be joined and no product is made. The need for an exact match on their adjacent ends makes it possible to discriminate between targets that differ by a single base pair. Multiplexing can be performed if one or more probe sets are added to the reaction.

OLA assays are used for many different applications in the clinical diagnostics laboratory [☞ "Multiplex Ligation Dependent Probe Amplification"].

Oligonucleotide priming

Oligonucleotide priming is a commonly used biochemical procedure in the molecular biology laboratory. It is an enzymatic method of labeling DNA probes with radioactively or otherwise tagged deoxyribonucleotides that are incorporated into newly made DNA molecules during the reaction. Once tagged, these DNA probes can be used as reporter molecules in subsequent tests to answer questions about DNA or RNA obtained from a patient specimen.

The word priming is used because the reaction depends upon DNA polymerase making new DNA strands from a template, which is first denatured. Such new DNA synthesis only begins if the DNA polymerase encounters a local region of double-strandedness to initiate DNA synthesis. The double-strandedness occurs through the addition to the reaction of short oligonucleotides (six to 10 base pairs in length). The sequence of these "oligos" is random; they find many complementary areas in the DNA that are to be labeled and bind at those locations. DNA polymerase in the reaction can then go about its business of making new DNA, incorporating into the newly made DNA strands tagged or labeled deoxyribonucleotides, also present in the reaction mixture. The tag or label may be radioactive, or be comprised of some other chemical entity that will allow subsequent detection of the hybrids formed between the DNA probe and its targets in the laboratory test to be performed. Priming between random oligos and the template strand is favored over simple sister strand re-association because of the vast molar excess of oligos used. In other words, simple competition favors priming due to overwhelmingly large numbers.

Since the sequence of the "oligos" is random, some refer to this biochemical reaction as random oligonucleotide priming. Originally, the "oligos" used were six base pairs in length, and this technique was known as random hexanucleotide priming [☞ "Autoradiograph;" "Chemiluminescence;" "DNA labeling;" "Nick translation;" "Oligonucleotide;" "Southern blot"].

Oncogene

Cancer-causing gene.

Many human genes are involved in controlling cell division and the rate of cellular growth. These genes have a normal, useful function and are called proto-oncogenes. When proto-oncogenes mutate, through any one of a number of mechanisms, they lose their ability to regulate cell growth and become cancer-causing oncogenes.

Examples of oncogenes include *abl*, *erb*B, *ras*, and *myc*. These and other oncogenes have been implicated in breast cancer, colon cancer, neuroblastoma (a childhood cancer), various kinds of leukemia and lymphoma, and other cancers. Discovery of the cellular origin of viral oncogenes, which led to the eventual elucidation of the role in cancer of proto-oncogenes and their mutated cousins, led to the awarding of the Nobel Prize in 1989 to J. Michael Bishop and Harold E. Varmus.

ORF

Open reading frame.

In the nucleotide sequence that comprises a gene are stretches of bases that will ultimately be translated into a protein. Each three successive bases, termed a triplet, codes for a corresponding amino acid (amino acids are the building blocks of proteins). Three triplets code for STOP signs. When the cellular machinery involved in elongating the protein chain (that is being translated from the DNA and RNA that code for it) encounters a triplet that signals STOP, the protein growth terminates and a mature (or prematurely terminated, mutant) protein has been generated.

An ORF, then, is a stretch of bases in DNA that *could* code for a protein because it has a specific START triplet [☞ "AUG"] and no STOP triplets (at least for a while until a reasonably sized protein can be generated from that stretch of bases). Previously unknown genes may be recognized in DNA sequences now available because computer analysis shows that there is an ORF (which *could* be a gene) that is then further investigated [☞ "Genetic code"].

Organelle

Think of this as a little organ. In the same way that organisms like *Homo sapiens* are made up of organs like the heart, lungs, brain, and kidneys, individual cells are made up of parts called organelles. Examples include the nucleus, cell membrane, mitochondria, and endoplasmic reticulum. Nuclei and mitochondria are DNA-containing organelles.

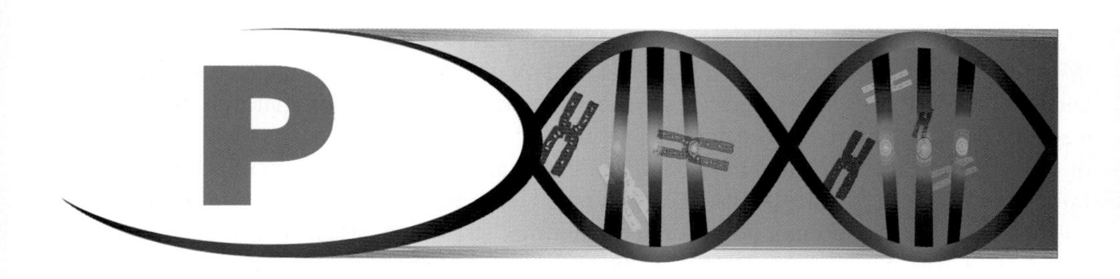

Paternity/profiling/identity/forensic testing by DNA

You know those bar codes that cashiers in supermarkets scan to figure out if you're buying a 99-cent jar of mustard or a $24 bottle of wine? DNA profiling is somewhat analogous. In the same way that the barcode reader allows a computer to apply software that discriminates the mustard (and the price) from the wine, one can use DNA patterns to individualize DNA specimens and thereby discriminate among them (and the individuals from which they were derived) from each other.

DNA specimens are individualized to determine paternity/non-paternity, or to search for matches between suspects and biological samples left at a crime scene. Medically, DNA specimens are individualized to ascertain if the bone marrow of the recipient of a bone marrow transplant is populated with cells from the donor-in other words, to assess the success or failure of bone marrow engraftment.

All individuals can be distinguished from each other at the DNA level. This is so because, with the possible exception of identical twins (who may differ from each other in copy number for particular sequences; ☞ "Copy number variation"), the DNA of any individual is different, at several different levels, from all other individuals. These so-called "genetic signatures" (the bar codes in our little analogy above) can be identified in the laboratory by Southern blots, PCR-based assays, and fragment analyses that exploit DNA polymorphisms (a fancy word for difference).

In general, polymorphisms refer to different forms of the same basic structure. At the DNA level, polymorphisms take different forms. The most significant one for identity testing is the different number of repeats in a repetitive DNA sequence. For example, the tetranucleotide sequence "AGCT" may be repeated 25 times in tandem in one person and 10 times in another. That polymorphism of 15 repeats (25–10) is detectable. Repeated sequences in DNA have been termed microsatellite or minisatellite DNA (or variable number tandem repeats [VNTRs]). The number of repeat units within microsatellite DNA is highly variable both within a single genetic locus [☞ "Locus"] and among different genetic loci.

Different probes and primers are available for several core sequences that exist within different hypervariable regions of the genome. When PCR-amplified DNA is hybridized with a probe specific for multilocus hypervariable sequences and subjected to electrophoresis, a complex pattern of bands appears on an autoradiogram or electrophoretogram (that looks not unlike a supermarket bar code). This pattern is unique for every individual. Alternatively, one may use probes specific for single locus hypervariable regions that are highly polymorphic (highly variable from person to person). The probability of two individuals having the same number of alleles in these highly polymorphic regions is quite low. Individualizing power becomes very great in this mode of analysis when additional single locus probes are used.

Micro- or mini-satellite repeats are also termed VNTRs or STRs (short tandem repeats). Within our genomes are segments of DNA that are variable in number (the VN in VNTR) and that reiterate a particular identical sequence within that segment of DNA, over and over, a so-called tandem repeat (the TR in VNTR). The number of repeat units within a micro- or mini-satellite repeat or VNTR is highly variable among individuals and can be determined, as described above, for purposes of identification. Microsatellite repeats are two to five bases in length; minisatellites about 10–100 base pairs in length.

TECHIE: Several PCR-based methods may be used to identify human DNA polymorphisms to individualize and identify them. Analysis for several different genes and genetic loci that exhibit polymorphisms among individuals include the HLA DQa locus; low density lipoprotein receptor (LDLR); glycophorin A (GYPA); hemoglobin G gammaglobulin (HBGG); D7S8; and group specific component (GC). Determination of "length polymorphisms" by PCR of amplified fragment length polymorphisms (AMP-FLPs) present in variable number tandem repeats (VNTRs) in the human genome may also be performed.

Half of an individual's DNA is inherited from each biological parent. The DNA testing (also known as "DNA fingerprinting") described can therefore be used to include or exclude an alleged father from that group of men that could be the biological father of a child. Similarly it could be used to establish maternity, although maternity is often assumed. Immigration questions also sometimes hinge on paternity and sometimes involve DNA fingerprinting. There are, of course, applications of identity testing to forensic and criminal investigation. An individual suspected of having committed a crime can be placed at the scene of the crime if the suspect's DNA fingerprint matches the genetic fingerprint obtained from DNA of hair, blood, semen, or other cellular samples found at the crime scene; the DNA fingerprint provides potentially valuable evidence for prosecution of that suspect. At the same time, a non-match between the DNA of the evidentiary material and the suspect may help exonerate the individual from the crime in question.

There are clinical applications of DNA fingerprinting. Bone marrow donors and recipients may be "fingerprinted" before a bone marrow transplant is performed. Such information may be used, post-transplant, to determine the success or failure of the engraftment procedure by analyzing the identity of the cells in the recipient's bone marrow. DNA fingerprinting may be important for genetic investigation to determine if twins, triplets, etc., are monozygotic (identical), or dizygotic (fraternal). Further, prenatal testing for genetic disease is often preceded by, or conducted simultaneously with, analysis of the gene in question of the parents of the unborn fetus. Obviously, it is crucial that analysis be conducted on the biological parents if it is to be informative with respect to the fetus, so DNA fingerprinting may be an important adjunct test.

DNA fingerprinting was first applied in two cases of rape/murder in the mid 1980s by Sir Alec Jeffreys in the United Kingdom. Sir Alec received the 2003 Association for Molecular Pathology Award for Excellence in Molecular Diagnostics.

Every year, an important international symposium on Human Identification is organized by Promega Corporation of Madison, Wisconsin. You can learn more by visiting Promega's website (www.promega.com and http://www.promega.com/applications/hmnid/). See Figures 25 and 26 [☞ "Microsatellite DNA"].

PBMC

Peripheral blood mononuclear cells (PBMC) are white blood cells, the lymphocytes and monocytes found in blood's cellular (as opposed to liquid) component.

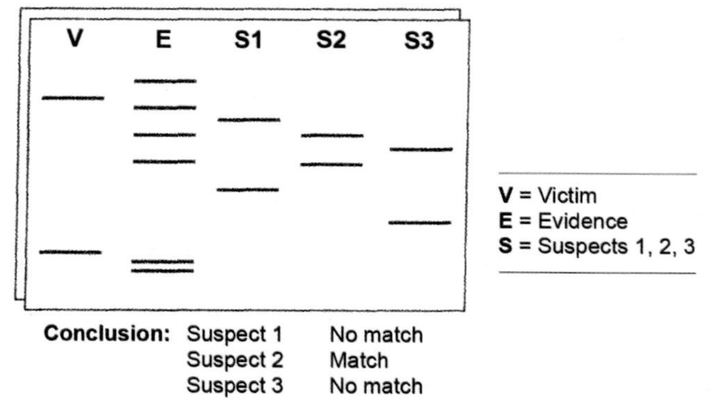

Figure 25. The figure shows a schematic of an electrophoretic gel separation of DNA fragments. The "fingerprint" or pattern of the victim's DNA is shown in the lane marked "V." The DNA extracted from the evidence (the "E" lane) shows a mixture of fragments. Only suspect #2 (S2) shows DNA in common with the evidence. In other words, suspect #2 matches the DNA fingerprint of the evidence and that match places that suspect at the crime scene. Suspects 1 and 3 are excluded.

Reprinted with permission from Promega Corporation, Madison, Wisconsin.

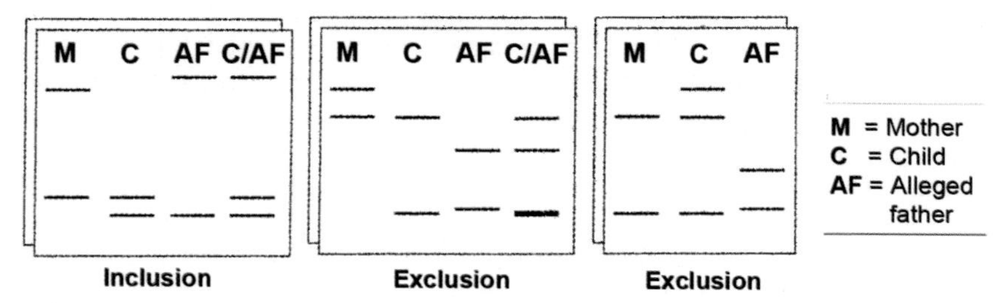

Figure 26. The figure shows a schematic of several electrophoretic gel separations of DNA fragments. The "fingerprints" or patterns of the mother's and child's DNA are shown in the lanes marked "M" and "C," respectively. "AF" stands for alleged father. In the "gel" marked "Inclusion," there is a match of one band between the AF and the child (the non-matching child band in this example matches a band in the "M" lane, demonstrating that one allele, or band, is inherited from the mother, while the second is inherited from the biological father). In the "Exclusion" examples, there are no bands in common between the child and the AF, demonstrating that the AF is excluded from that group of men who could be the biological father. In each case the "C/AF" lanes represent a mixture of child and alleged father DNAs run in the same gel lane. In the middle gel the broadness of the bottom band in the C/AF lane demonstrates that there are actually two bands there that may or may not be appreciated by simple visual inspection of the "C" and "AF" lanes separately.

Reprinted with permission from Promega Corporation, Madison, Wisconsin.

These cells are often used as the source for isolation of nucleic acids from blood specimens. PBMCs can be isolated directly from whole blood or from the buffy coat. The buffy coat is the WBC portion of blood that is isolated using centrifugation to separate the different blood components. After appropriate centrifugation, the buffy coat is in the middle of the tube, with the red cell component on the bottom and the plasma component on the top; the plasma can be aspirated away (or saved) and the buffy coat isolated for subsequent molecular diagnostic work.

PCR (and RT-PCR)

Polymerase chain reaction (PCR; see Figures 27 and 28).

You've heard the cliché "looking for a needle in a haystack." Well, PCR is a biochemical reaction that finds the needle, and then generates a haystack full of those needles in a very specific way. PCR is based on how DNA is naturally replicated in the cell. In the cell, the DNA strand to be replicated is called the template. In order for repliction (a normal part of cell division) to occur, (1) the double-stranded DNA molecule is separated; (2) a short naturally occurring primer binds to the region being replicated (this is an oversimplification but we don't want to write a nucleic acid biochemistry text book); and (3) the enzyme, DNA polymerase, then adds the appropriate nucleotides based on the template sequence.

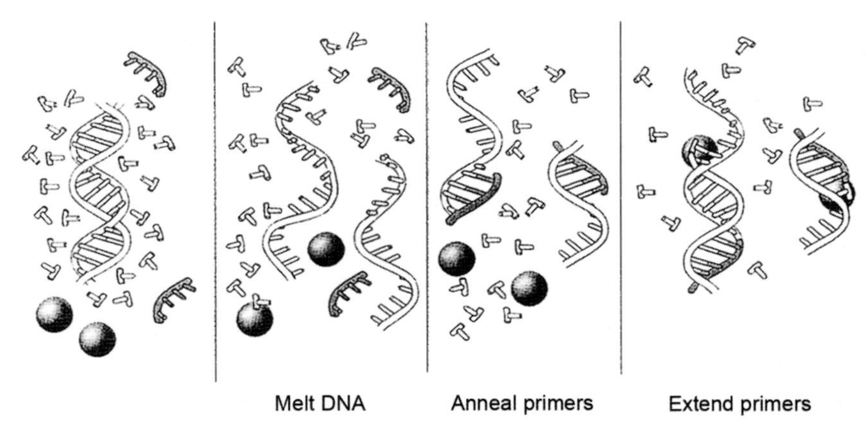

Figure 27. PCR starts with sample or template DNA, primers, nucleotides, and polymerase. The first step melts the DNA to single strands, providing access so the primers can anneal in the next step. Then polymerase extends the primers. Each repetition of this cycle doubles the amount of target sequences.

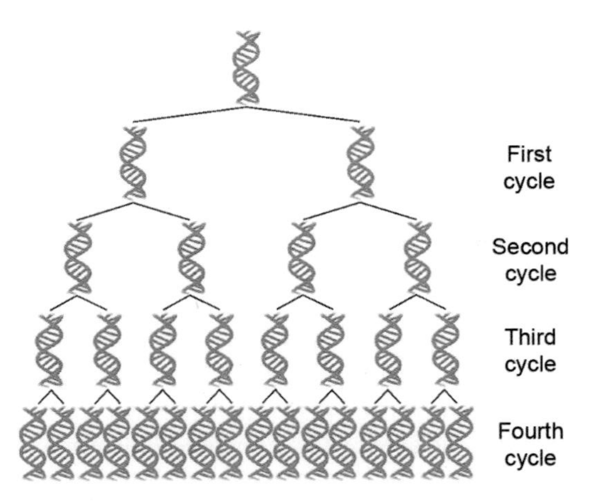

Figure 28. Every PCR cycle doubles the amount of target sequence. DNA synthesized in one cycle serves as a template in the next cycle. The four cycles depicted here produce 2^4, or 16, copies.

In the laboratory, we use the same strategy. In a molecular diagnostics laboratory-based clinical test, the DNA template is isolated from a patient specimen (it may be a test for detection or quantification of a pathogen, in which case that bacterium's or virus's genome is the target). In PCR, primers bind to the template region (target of interest) to be replicated (a million-fold or more; much more than would naturally occur in a cell) and the same enzyme, DNA polymerase, is used to add nucleotides in the same manner as in normal cellular replication.

Here's how PCR works:

Patient DNA is purified and heated to a temperature near boiling (~94–95 °C) in a reaction tube that contains all the necessary ingredients for PCR. The high temperature makes the naturally double-stranded DNA single stranded; this is the **denaturation** step. The reaction temperature is then lowered to 30–65 °C, and the oligonucleotide primers present in the reaction mixture find the bases in the patient DNA to which they are complementary, and bind. This is the **annealing** step and creates a local region of double-strandedness. Two additional components, DNA polymerase and the building blocks of DNA, the deoxyribonucleotide triphosphates (dNTPs for short), then come into play. Using target DNA, e.g., suspected mutation, viral gene, as a template, DNA polymerase adds the dNTPs to the bound primer, creating a new complementary DNA strand. This is called the **extension** step and occurs when the temperature of the reaction is changed to ~65–75 °C. The three steps taken together, **denaturation, annealing,** and **extension,** constitute one PCR cycle; one cycle is typically repeated 25–40 times or more during PCR to amplify the target of interest in the molecular diagnostics laboratory.

Each cycle doubles the amount of DNA from the previous cycle. So, for each cycle, one double-stranded DNA molecule is denatured into two single-stranded molecules, primers anneal to each strand, and DNA polymerase, by adding dNTPs, creates two new double-stranded DNA molecules; one target becomes two. In the second cycle, upon repeating the process, two targets become four; in subsequent cycles the target number increases to eight, 16, 32, etc. After about 30 cycles, a billion-fold increase in the amount of starting DNA has been generated. We start with a needle in a haystack (the DNA target of interest); upon the completion of PCR, a haystack full of specific needles has been generated by this specific copying mechanism called PCR, something of a molecular photocopying machine.

The astute reader will realize that proteins, such as the DNA polymerase used in PCR, can't withstand temperatures as high as 94 °C. In fact, most proteins become irreversibly denatured (unraveled and destroyed) and do not function at such high temperatures (nucleic acids can withstand such temperatures). An important technical advance for PCR and one that facilitated its automation came with the realization that there are bacteria that normally carry on the business of life in hot springs, like those in Yellowstone National Park. One bacterial species, *Thermus aquaticus*, lives in such springs at temperatures of 75 °C; DNA polymerase purified from this bacterium functions at temperatures over 90 °C. This DNA polymerase, termed *Taq* polymerase after the bacterium from which it is purified, is the workhorse of PCR and has been a significant factor in the adaptation of PCR in the clinical molecular pathology laboratory.

The specificity of PCR is dictated by the sequence of the primers used in the reaction. The DNA sequence of interest must be known in order to design primers that bind to regions on either side of the target, thereby bracketing it and defining the region to be amplified. The primers need to be long enough (15–25 bases) so that they bind only to the specific regions of interest and not randomly throughout the genome. Once specific primer binding occurs, PCR functions efficiently to create more and more amplified target so that medical technologists can analyze it in the laboratory. Often PCR copies are several hundred base pairs in length and used in subsequent detection reactions; in so-called real-time PCR, amplification and detection are essentially simultaneous. Real-time PCR represents a great advance in 21st century molecular diagnostics.

PCR is not limited to examining only the human DNA in a specimen. If an infectious agent is suspected of causing disease, PCR can be used to detect the presence of a DNA target specific for the microorganism in question. The DNA from the infectious agent would have been purified at the same time as the DNA from that particular clinical specimen, for example, a blood specimen. If detected, then the PCR test for that particular bacterial agent or virus is positive.

Some viruses are RNA viruses by nature, for example, hepatitis C virus and human immunodeficiency virus (HIV). RNA is not a suitable starting material for PCR; however, the addition of one extra step in the PCR process solves that problem. Reverse transcriptase (RT) is an enzyme that naturally synthesizes DNA from RNA as starting material. RNA plus the RT enzyme yields DNA that is designated cDNA (c for complementary). The cDNA generated by RT is a perfectly suitable DNA molecule to participate in subsequent PCR. Think of this as RNA-PCR; it is abbreviated RT-PCR (for Reverse Transcriptase Polymerase Chain Reaction). An enzyme called *Tth* polymerase combines the activities of reverse transcriptase and DNA polymerase, and is used in most laboratory RT-PCR reactions. *Tth* polymerase is a thermostable (stable at high temperatures) enzyme from the bacterial species, *Thermus thermophilus*, hence the name.

The genesis of the idea for PCR occurred to the scientist Dr. Kary Mullis in the early 1980s on an evening drive through Northern California's Redwood country (you can read a detailed description of the story in his Nobel laureate lecture (http://www.nobel.se/chemistry/laureates/1993/mullis-lecture.html). Mullis made the reaction work in the laboratories of Cetus Corporation where he was employed. Since the advantage of *Taq* polymerase's thermostable nature was not yet contemplated, new DNA polymerase had to be added at every cycle because the near-boiling temperatures needed for DNA denaturation irreversibly denatured ordinary DNA polymerase. In addition, early PCR, including the work of Mullis, depended on dedicated scientists sitting by the tubes involved in PCR with a stopwatch and several water baths, set to different temperatures. The reaction tubes were transferred manually among three different water baths, each set to the temperature needed for each step in the cycle—and there were up to 40 cycles. This made the entire PCR process very labor intensive and tedious. Today, and for the last 20 years or so, instruments called thermal cyclers facilitate rapid temperatures cycling, allowing PCR to be performed in rapid, automated fashion.

Throughout the late 1980s and early 1990s, PCR took the scientific community by storm. PCR is probably the single most important discovery that led to the field of molecular pathology (keeping in mind of course that Southern blotting, nucleic acid extraction, DNA sequencing, and restriction endonuclease digestion are all important tools whose utilities are based on the fundamental elucidation of the double helical structure of DNA). PCR is largely responsible for molecular pathology becoming a mainstream part of modern-day clinical pathology laboratories.

In addition to traditional PCR and RT-PCR, many variations exist. Some variations are used in clinical laboratories, and others are strictly used in research settings. Table 6 lists

Table 6. PCR

PCR Variation	Principle Behind Reaction	Possible Applications
Two-cycle PCR	The annealing and extension steps are combined.	"Need for speed;" "stat" test result needed
Asymmetric PCR	Excess concentration of one primer over the other.	To amplify one strand preferentially over the other; when interrogating a specific allele.
Multiplex PCR	More than one primer pair in a single reaction.	To amplify more than one target in the same reaction, e.g., detecting both *Chlamydia trachomatis* and *Neisseria gonorrhoeae* from the same specimen.
Nested PCR	Two PCR reactions and two primer sets; first reaction uses one primer pair, second reaction uses primer pair that anneals inside ("nested" within) first.	Provides increased sensitivity; useful for small amounts of starting nucleic acid, e.g., small specimen volume.
Real-Time PCR	Fluorescent molecules used to detect product as it is made, i.e., simultaneous amplification and detection.	Useful in tests requiring rapid turn around time, e.g., patient in the ER, result needed to prescribe correct antibiotic.
Touchdown PCR (no football jokes, sorry)	Decreases annealing temperature in 0.5 °C increments each cycle until a specific target temperature is reached.	When target sequence is variable, e.g., some RNA viruses.

some of the most popular variations, what makes them unique, and potential applications. Beyond the dramatic role of conventional PCR in transforming molecular diagnostics, real-time PCR has revolutionized the field yet again; it has had such a major impact that though it is listed in the table, please see a more thorough description under its own heading [☞ "Real-time PCR"].

The rights to PCR were purchased by Hoffman-LaRoche for $300,000,000 (think about that and the value of that number, a lot of money even when this deal was executed 20 years ago). As you might imagine when such sums are involved, the transaction precipitated litigation. Speaking of litigation, we have heard Dr. Mullis advise that if you ever have a choice between putting your name next to "Patent Inventor" or Patent Assignee," choose the latter; it is much more lucrative.

PCR is covered by patents owned by Hoffman-LaRoche. If one performs an Internet search for the term, "PCR," many interesting sites, including some with animation, are listed [☞ "cDNA;" "Denature;" "Oligonucleotide;" "Primer;" "Real-time PCR;" "Retroviruses;" "Reverse transcriptase"].

PCR "wannabes"

It was obvious early on that PCR was a powerful technology. Not only did it have myriad research applications but it was clear that PCR had the potential to revolutionize medical diagnostics testing (indeed, since realized). Today, many PCR tests are FDA-approved or cleared for diagnostic use [☞ Table 3 under "FDA"]; many more are somewhere in the developmental/regulatory pipeline. This situation is another example of where medicine, law, and money (markets) intersect. PCR is medically important, is covered by patents owned by Hoffman-LaRoche, and has a tremendous real and potential marketplace. Diagnostics companies, realizing this, set their research scientists on the task of developing biochemical reactions similar to PCR—"PCR" wannabes". Many examples exist from many different companies and they are innovative and beneficial. Some involve complicated biochemistry while others use more straightforward biochemistry to accomplish the same end result: in vitro nucleic acid amplification (making a haystack full of specific needles in the test tube). Some are listed in the FDA table (Table 3) alluded to above.

Products in the marketplace include Transcription-Mediated Amplification (TMA), marketed by Gen-Probe, Inc., in San Diego, CA; branched DNA amplification [☞ "bDNA"], marketed by Siemens Healthcare Diagnostics in Deerfield, IL; Nucleic Acid Sequence Based Amplification [☞ "NASBA;" name changed to product line, NucliSens], marketed by bioMerieux, which has facilities in Europe and N.C.; and Linked Linear Amplification (LLA), marketed by Bio-Rad of Hercules, CA. Note that LLA generates linear, not exponential, amplification. Linear, as opposed to exponential, amplification is sufficient to answer some analytical questions, particularly when genotyping individuals where sensitivity is less of an issue than in molecular infectious disease testing, where they may be very small numbers of pathogens in the patient specimen.

Peptide nucleic acid (PNA)

A lot of very creative people have gone into the field of molecular biology. The development of peptide nucleic acids (PNAs) is one example of this creativity.

Natural DNA is a double-stranded molecule consisting of nucleotide bases attached to a negatively charged phosphate-sugar backbone. The negative charge of each of the backbones imparts a natural repulsion that facilitates separation of the two strands during replication and transcription. This natural repulsion, while great for natural processes, is not ideal when strong binding is required, i.e., when binding short probes. Here's where the creativity comes into play: in order to eliminate the natural repulsion of the two DNA strands, some envisioned the creation of a new molecule that was uncharged. In order to do this, three requirements needed to be met: (1) the new molecule needed to form a chain; (2) the nucleotide bases needed to attach to it and; (3) the molecule had to be roughly the size of a normal chain to allow base-pairing. The result was Peptide Nucleic Acid: PNA.

PNAs are synthetic molecules where the phosphate-sugar backbone normally present in DNA is replaced with a glycine backbone. Glycine is a small amino acid (this is where the name peptide in PNA comes from) that does not have a charge. The nucleotide bases can be linked to the glycine backbone to make a short strand, allowing it to function more or less the way natural DNA functions. PNAs bind with DNA, RNA or other PNAs, forming strong bonds. PNAs are attractive alternative molecules for use in diagnostic and research assays.

Personalized medicine

Laboratory medicine already is personalized medicine; we test Patient A's cholesterol level or Patient B's HIV viral load. Laboratory tests are done in batches for the sake of efficiency but the results are highly individualized and reported as such into a patient's medical record. What's really meant by the hot catchphrase "personalized medicine" is a more serious attempt at using genomics and molecular diagnostics to tailor therapies, i.e., to personalize therapies beyond the typical "trial and error" approach.

The "trial and error" approach has worked well because pharmaceutical companies design and manufacture drugs that address the overwhelming majority of patients with a particular set of symptoms; to do otherwise would be economically unsound. And because drugs work well on so many patients only a relative few suffer adverse reactions or don't receive relief. Personalized medicine is an attempt to remove the small amount of "error" in the system and use genomics in advance to learn if a patient is a suitable candidate for a particular drug, i.e.,will the patient tolerate the drug, or suffer an adverse reaction? Will the patient respond to the usual dosage? It would be ideal to know these things in advance of drug administration. Learn more at www.personalizedmedicinecoalition.org.

Pharmacogenomics

Researchers have come to find that an individual's response to therapeutic drugs is partially a function of that individual's genetic makeup. Given that our genomes define all characteristics of an individual, to a greater or lesser extent, depending on environmental factors, it's actually not a surprising finding that drug response is also genetically defined. Just as allergic reaction to ragweed pollen is genetically determined (it bothers our spouses but not us), so, too, is the ability or inability to metabolize a drug. Thus, a particular drug may prove effective in one individual while the same drug is ineffective or, worse, causes a severe adverse effect in a second.

Today, when we present to our physicians with a given set of symptoms, the physician usually prescribes a particular drug that clinical medicine has learned is valuable in the treatment of a specific set of symptoms. Often, the drug works to one degree or another. Indeed, the reason the drug reached the market is that it was shown in clinical trials to be effective in the vast majority of individuals suffering from a given, defined set of symptoms.

Sometimes, however, drug therapy fails. Failure may be caused by genetic variation in the ability to effectively metabolize that drug; for example, so-called ultra-rapid metabolizers may so aggressively metabolize a drug that a therapeutic level is not achieved with the normal dose, i.e., the dose designed for normal (called extensive) metabolizers. Alternatively, genetically controlled drug metabolism may be so compromised in some individuals (so-called poor metabolizers) that the normal dose generates no relief in the patient and the drug builds up to potentially toxic levels after repeated doses.

The day is at hand and laboratories are now able to do studies on individuals for particular drug response genes for specific drugs (Table 7; a full listing is available at: http://medicine.iupui.edu/flockhart/table.htm). Based on such tests' results, physicians who order these tests may prescribe in a tailored, specific fashion, the right drug at the right dose perfectly suited to that individual, thereby maximizing the chances that the drug therapy will be successful. Think of it as rational drug prescription rather than "trial and error" (keeping in mind that "trial and error" very often works, based on rich medical experience, but wouldn't it be nice to increase the percentage of effective drug treatment while decreasing the percentage of adverse drug reactions?).

As the word "pharmacogenomics" implies, this field is a synthesis of genomic investigation and pharmaceuticals. Many drug companies are investing huge sums of money in

Table 7. Drug Response Genes

Drug (Generic Name)/Indication	PGx Test or Correlation	Comment
Herceptin (Trastuzumab)/breast cancer	HercepTest and PathVysion test	Amplification of HER-2/*neu* gene determine if patient is an appropriate candidate for drug.
Gleevec (Imatinib mesylate)/chronic myelogenous leukemia (CML) and gastrointestinal stromal tumors (GIST)	CML: *bcr/abl* translocation (PCR, FISH, choromosomal translocations); GIST: *kit* mutation and/or overexpression analysis	T315I mutation may render leukemic cells resistant to Gleevec's (second-generation TK inhibitors, dasatinib and nilotinib) anti-TK activity, which checks leukemic cell growth.
Camptosar (Irinotecan)/colon cancer	Invader test for uridine diphosphate glucoronosyltransferase 1A1 (UGT1A1)	Topoisomerase inhibitor; specific polymorphisms in UGT1A1 may cause improper drug metabolism and adverse effects (severe diarrhea and neutropenia).
Iressa (Gefitinib)/advanced non-small cell lung cancer	Test for expression of EGFR to identify drug candidates	Drug inhibits tyrosine kinase activity associated with EGFR.
Tarceva (Erlotinib)/non-small-cell lung and pancreatic cancer; some hematologic disorders (off-label)	Solid tumors: IHC test for overexpression of EGFR; hematologic disorders: test for *JAK2* mutation	EGFR and *JAK2* overexpression are associated with oncogenesis.
Avastin (Bevacizumab)/metastatic colorectal cancer and non-small-cell lung cancer	Detection of VEGF overexpression may or may not be useful	Drug inhibits tumors from developing a blood supply.
Ziagen (Abacavir)/anti-HIV retroviral	HLA-B*5701	Gene variant highly associated with hypersensitivity reactions to drug.
Coumadin (Warfarin)/anticoagulant used to prevent blood clots and pulmonary embolism	Tests for cyp450 2C9 and VKOR1 polymorphisms	Cyp450 variants, CYP2C9*2 and *3, metabolize warfarin inefficiently; VKOR1 polymorphisms are responsible for variations in warfarin metabolism.
Tegrtol, etc. (Carbamazepine)/anticonvulsant approved for mania symptoms in BPD	HLA-B*1502	FDA recommends testing for gene variant in patients of Asian ancestry to avoid risk of dangerous or fatal skin reactions (Stevens Johnson syndrome; toxic epidermal necrolysis).
Tamoxifen	Cyp450 2D6 variants	2D6 poor metabolizers and those receiving 2D6 inhibitors, e.g., fluoxetine, fail Tamoxifen therapy.
Psychotropic medications, e.g., Amitriptyline, Fluoxetine, Nortriptyline, Paroxetine, Venlafaxine, Haloperidol, Risperdone/depression, and antipsychotics	Variations in cyp450 alleles, including 2D6 & 2C19	Many tests for variants exist in the marketplace. Learn more at http://www.ahrq.gov and search for cyp450.

BPD, bipolar disorder; EGFR, epidermal growth factor receptor; FISH, fluorescent in situ hybridization; IHC, immunohistochemistry; PCR, polymerase chain reaction; PGx, pharmacogenomics; TK, tyrosine kinase; VEGF, vascular endothelial growth factor; VKOR1, vitamin K epoxide reductase.

NOTE: An excellent review on the pharmacogenomics of Cancer-Candidate genes is available (http://www.bcbs.com/betterknowledge/tec/vols/22/22_05.pdf).

this field in hopes that it will help shave years off the drug discovery and approval process, bringing potentially lucrative drugs to market that much sooner. Furthermore, pharmacogenomic testing has the potential to "rescue" drugs that have failed in trials by identifying in advance patients who may suffer an adverse drug reaction.

The best possible way to take full advantage of pharmacogenomics would be at the point-of-care, in the physician's office before the prescription is written. Consider the day when a buccal swab (scraping of the inside of the cheek to harvest a few DNA-containing cells) or blood specimen is drawn in the physician's office, the DNA automatically extracted and then analyzed in a point-of-care instrument that assesses the patient's DNA for key genetic markers, all in 30–60 minutes. These markers in various relevant genes will be chosen because they inform the physician that, for example, drug A is a better choice for this patient than drug B because this patient is incapable of metabolizing drug B into its active form and it would therefore have no therapeutic effect. At the height of convenience, imagine a computer or web-based link between the pharmacy and the physician's office such that the results are available when the patient reaches the pharmacy and no time is spent waiting in the physician's office for the result. These are not far-fetched scenarios.

Information relevant in pharmacogenomics is obtained by experimentation and analysis. Such analysis was traditionally done in a living organism (in vivo) or in a test tube or experimental vessel (in vitro). The nature of pharmacogenomics, which relies heavily on informatics and the ability to analyze relevant DNA sequences in computer databases, has led to a new way to describe experimentation: in silico.

Phenotype

Phenotype is the manifestation of a particular genetic makeup, i.e., the genotype, of an organism. Examples of phenotypes include blue eye color, affected with cystic fibrosis, inability to metabolize morphine, and maleness "Genotype;" "Allele"].

Plasma

Plasma is the fluid portion of blood. In preparation for clinical laboratory testing, blood is often collected in tubes containing an additive to prevent clotting, a so-called anticoagulant. Examples of anticoagulants include heparin or EDTA. The collected blood may then be subjected to centrifugation to force the solid parts of the blood (the cells) to the bottom of the tube; the liquid portion that remains is plasma. The cells are a good source of DNA for molecular diagnostic testing. Plasma is a good source of testing for analytes present in the liquid portion of blood, for example, proteins like antibodies in immunologically based tests.

Blood may also be collected in the absence of an anticoagulant and allowed to clot in the collection tube. If this specimen is then subjected to centrifugation to force the solid parts (of the clot) to the bottom of the tube, the liquid portion that remains is called serum. The difference between serum and plasma is that plasma retains those proteins responsible for clotting (which are inhibited from acting due to the presence of the anticoagulant) [☞ "Serum].

Plasmid

A plasmid is a circular piece of DNA that exists in bacteria as an extrachromosomal element. That means it exists outside and separate from the single chromosome of a bacterial cell.

Plasmids are smaller than the bacterial chromosome and many replicate autonomously, that is, independently of the rest of the DNA in the bacterium. Molecular biologists have learned how to use plasmids for cloning; plasmids are commonly used cloning vectors. To clone DNA, a fragment of DNA of interest, e.g., a gene, is isolated and inserted into a plasmid. This so-called recombinant piece of DNA is then introduced back into a bacterial cell where the plasmid will thrive and replicate. As it replicates, more and more of the inserted DNA is also made. This is an example of genetic engineering [☞ "Clone;" "Genetic engineering;" "BACs;" "HACs;" "YACs;" "Vector"].

Plasmids are also responsible for some types of antibiotic resistance. Organisms such as *Staphylococcus aureus* can harbor plasmids that make them resistant to virtually all antibiotics available. Since the plasmids are passed on to "daughter" bacterial cells, all subsequent generations will also be resistant to the antibiotics. As if that isn't bad enough, *S. aureus* strains that have the resistance plasmid can pass it to *S. aureus* strains that do not carry it.

Molecular pathologists exploit the plasmid present in the bacterium *Chlamydia trachomatis.* This microorganism causes the most common sexually transmitted disease in the United States (~4,000,000 cases/year; approximately 3,000,000 cases are reported annually in Europe). There are important health consequences to untreated *C. trachomatis* infection, including pelvic inflammatory disease, infertility, and more. Multiple tests are available for the detection of this microorganism in patient specimens (cervical swabs, urethral swabs, and urine) that work by amplifying *C. trachomatis*-specific DNA sequences (see Table 3 under "FDA" entry on page 53). The target in some of these tests is DNA in a bacterial plasmid that is present at about 10–100 copies per bacterial cell. By using this plasmid DNA as the target, the "deck is stacked" in favor of sensitive detection because the target has been naturally amplified by the bacteria. The plasmid DNA is therefore a better target of detection than a portion of the bacterial chromosome that is present at only one copy per bacterial cell.

Ploidy

Ploidy refers to the number of chromosomes in a cell. In every normal human somatic cell (except red blood cells; remember that sex cells, or gametes, are by definition not somatic cells), 23 pairs of homologous (one from each pair derived from Mom; one from Dad) chromosomes exist; because of the fact that there's a pair, these cells are said to be *di*ploid (*di* for two). Gametes (ova and sperm cells), on the other hand, have one set of chromosomes that are not paired; 23, rather than 46 chromosomes; these are called *hap*loid (*hap* for one). Variations on this theme include triploidy (3 sets or 69 chromosomes); tetraploidy (4 sets); and aneuploidy (a number of chromosomes, in humans, not divisible by 23—for example, Down Syndrome patients have three copies of chromosome 21 for a total of 47, not 46).

Polymerase

An enzyme (protein) whose job is to polymerize or make more of its substrate (target) (See Figure 29).

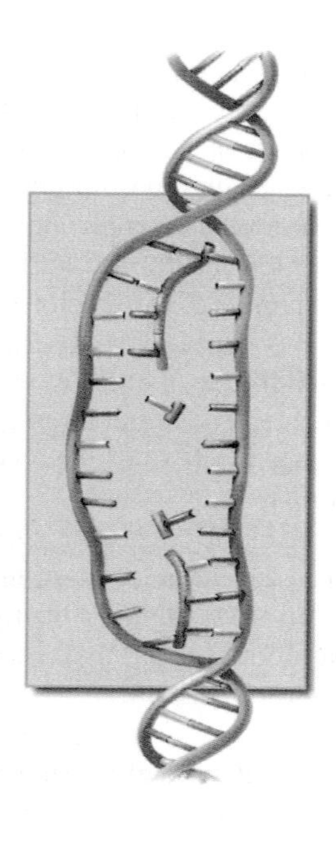

Figure 29. DNA polymerase can't start a DNA strand from scratch, but can only extend existing strands. Thus, synthesis takes off from short primers, so-called because they prime the process.

DNA polymerase I is the enzyme involved in making more DNA; RNA polymerase naturally synthesizes RNA. By the way, these, as all proteins, are also encoded by their respective genes within the DNA of the organism, whether that organism be a human, animal, plant, bacterium or virus [☞ "Expression"].

There are several classes of polymerases like those that make DNA or RNA, as described above. This subdivision, however, goes a bit further. A nucleic acid polymerase acts using another nucleic acid as the template to direct that synthesis. The template can be DNA or RNA, depending on the enzyme. There are four general kinds of polymerases, then:

1. DNA-Dependent DNA Polymerase (DDDP; catalyzes the synthesis of DNA using DNA as the template)
2. DNA-Dependent RNA Polymerase (DDRP; catalyzes the synthesis of RNA using DNA as the template)
3. RNA-Dependent RNA Polymerase (RDRP; catalyzes the synthesis of RNA using RNA as the template)
4. RNA-Dependent DNA Polymerase (RDDP; also known as Reverse Transcriptase; catalyzes the synthesis of DNA using RNA as the template)

Polony

[☞ "DNA sequencing"]

Primer

Think about painting your room or your house. There are some surfaces that paint won't stick to very well, so you first cover the surface with a primer. Once the primer adheres and dries, you can apply the paint successfully and stand back and admire your excellent handiwork.

This is a good analogy for the biochemistry of primers as they relate to DNA molecules. An enzyme (protein) called DNA polymerase (paint brush) makes more DNA when it has the raw materials (paint; dNTPs) to do so. DNA polymerase makes more DNA using unwound DNA as a template. This unwound DNA is single-stranded. So if that strand reads ATTAGCC, it directs the synthesis of a new strand complementary to it: TAATCGG [☞ "Complementary strands of DNA"].

But DNA polymerase won't work without a small section of double-stranded DNA to initiate or "prime" new DNA synthesis (the "paint" won't stick to the wall). In the polymerase chain reaction (PCR), which occurs in vitro, and during DNA replication that occurs in vivo, small segments of DNA of a defined length called primers are added to prime the site of initiation of DNA synthesis by DNA polymerase. These primers are oligonucleotides (oligo means few) [☞ "In vitro;" "In vivo;" "Oligonucleotide"].

When designing oligonucleotide primers in silico for use in the laboratory, one needs to be careful to avoid creating a sequence where one end of the oligo happens to share complementarity with the other end of the sequence. If this happens, the two ends will find each other, base pair very happily to each other, and form what is known in the field as "hairpins."

Another type of primer design that should be avoided is complementarity between the 3'-ends of the primer pair. If this happens, the primers will anneal to each other and the polymerase will extend each of the primers using the other primer as the template. This is called a primer-dimer.

Good oligonucleotide probes and primers should not form hairpins or primer-dimers or exhibit complementarity in the middle part of their sequences if they are to be used successfully [☞ "Hairpins;" "Primer-dimer;" "In silico"].

Primer-dimer

A primer-dimer is an undesirable, short, amplified product that is the result, during PCR, of the sense primer binding to the anti-sense primer at their 3'-ends instead of to the desired target sequence. This unwanted dimer is then extended by DNA polymerase resulting in a short duplex.

Come again?? Ok, let's say we are designing a laboratory test to amplify a short sequence. We design an upper primer (the sense primer) and a lower primer (the anti-sense primer). When PCR is completed, we expect to observe a 200 base pair (bp) DNA fragment but instead detect a 40 bp fragment. What happened? A primer-dimer was formed. What this means is that the primers are complementary to each other on their ends so they bind to each other instead of the intended target. Because DNA extends from the 3'-end, the enzyme is able to extend from the primer, using the other primer as a template. The following primer pair is an example of primer-dimer formation:

Step 1. Primer binding

```
5'-ATACTGCGAATATTCGCAGAGTACGT- 3' →
                         ||||
                    ← 3'-TGCAACTGATGCCATGACTAGAATCAT-5'
```

(bold: ACGT in the upper strand, TGCA in the lower strand)

(Note the bold nucleotides' complementarity.)

Step 2. DNA polymerase extension of the two primers in the direction of the arrows, creating a short product

(Note that the same nucleotides are bolded for reference.)

Primer-dimers along with other primer binding configurations such as hairpins must be avoided in designing successful assays. Many computer programs are available that will not only design the primers but will also check for potential unwanted primer configurations. These programs have proven invaluable for designing successful assays [☞ "Hairpins;" "Primer"].

Primer extension assay

TECHIE: If one would like to identify whether or not a single mutation is present, one could choose to use a primer extension assay. These assays rely on an exact match between the 3'-end of the primer and the template.

Let's say we have been asked to investigate if a patient has a specific mutation that causes abnormal blood clotting (tests are initiated by physicians' requests). The known mutated base is an adenine (A), where guanine (G) is normally present. We design a primer that has the complement of the A (thymine; T) on the 3'-end (recall that DNA polymerase extends off the 3' end of DNA). During the annealing step (binding of the primer to the target), the primer will bind both sequences, but only the sequence with the exact match will be extended by the polymerase. Since the polymerase is not able to extend the primer unless it is firmly attached to the template at the 3'-end, the primer will not be extended if it binds to a target containing the G base, therefore no product is made. With observation and the proper controls, this method can be used to genotype an individual for a particular mutation [☞ "PCR;" "Allele-specific primer extension"].

Proband

The Eagles, The Rolling Stones, The Glenn Miller Band, The Dave Matthews Band—these are all Pro Bands. A proband in genetic terms is the individual in a family afflicted with genetic disease whose study brought that disease within the family to light. Genetically speaking, the extended family in which this particular heritable disease is documented is called a kindred or pedigree. The proband may be thought of as the index case. Some also call the proband the propositus.

Probe

In molecular biology, a probe is a relatively small piece of DNA used to find another, related, piece of DNA.

In nucleic acid hybridization, a DNA probe must be labeled (radioactively or non-radioactively). The label enables the probe to "report" back that it "found" the DNA the molecular pathologist sought, for example, a genetic mutation and the sequence surrounding it in patient DNA subjected to a laboratory test. Based on the tag used to label the DNA, different methods are employed to detect hybridization and answer questions about the patient's DNA that are relevant to the diagnostic issue at hand. The shortest useful probe is about 20 bases long and is known as an oligonucleotide probe. This is about the length of probes used in real-time PCR assays. Probes can be much longer in molecular pathology laboratory tests like the Southern blot.

RNA molecules may also be used as probes and are termed "riboprobes" [☞ "Autoradiograph; "Chemiluminescence;" "Complementary strands of DNA;" "DNA labeling;" "Hybridization;" "Real-time PCR;" "Southern blot"].

Prognostic genomic testing for breast cancer

Several such tests exist in the marketplace from companies such as Agendia, AviaraDx, Dako, and Genomic Health (two are FDA-approved at this writing in spring 2008); there are others and undoubtedly more will be on the market by the time this book is published and as it ages on your shelf.

This class of tests has the potential to assist the treating oncologist and the breast cancer patient with the sometimes difficult decision of adjuvant chemotherapy for some cases of breast cancer that may be well-managed and/or cured with lumpectomy alone (with possible local irradiation). Small or tiny tumors that have not spread have a high likelihood of cure with such treatment and often the decision about adjuvant chemotherapy to further reduce the possibility of tumor recurrence, even if infinitesimally, is a difficult one. Chemotherapy is unpleasant and expensive; if it adds no benefit, why pursue it? Genomic-based tests that assess the gene expression profile of panels of genes exist; the data are then uploaded to a software-based algorithm that generates a recurrence score and/or information that can be used to predict tumor recurrence. If the risk of recurrence is very low, then a patient and her oncologist may be comforted in a collaborative decision to forego chemotherapy for small node, negative tumors. If the risk is high, then at least the disadvantages of chemotherapy can be "tolerated" knowing that it is likely to be of some benefit.

As with most things in life, if the likelihood of recurrence as determined by one of these tests is moderate or "in the gray zone," then the decision is of course more difficult. Patients are likely in the best position possible by gaining as much information as they can and working very closely with their oncologist in making such important and personal decisions.

Promoter

Just as prizefighters need the very best promoters to advance their pugilistic careers, so, too, do genes need promoters. Within the realm of molecular biology, promoter refers to a stretch of bases just upstream (in front) of the start of a gene.

Gene expression begins with transcription of DNA into RNA and this begins at the so-called transcription initiation site (TIS). An enzyme, RNA polymerase, whose job is to synthesize an RNA transcript from a DNA template, binds to the TIS, within or juxtaposed to the promoter. Once bound, RNA polymerase can carry out its task.

A protein called dystrophin is a common protein component of muscle. Mutations in dystrophin lead to the general class of diseases called dystrophinopathies, the most infamous of which is muscular dystrophy (MD). The most severe form of MD, Duchenne MD (DMD), is caused by mutations that severely disrupt the gene's coding sequence, leading to a truncated protein or no protein expression whatsoever. A milder form of MD (Becker's MD) results from a mutation that leads to a longer, more highly functional protein than in DMD, but still less functional than normal dystrophin. A third kind of mutation in the dystrophin gene was found in cardiac muscle. This particular mutation leads to a heart-specific dystrophinopathy called X-linked cardiomyopathy, a generally fatal disease. The mutation is in the promoter for the gene such that the dystrophin gene is simply not transcribed into RNA in cardiac muscle.

In cancer, promoters are often a site of mutation or site of the results of mutation leading to aberrant gene expression and disturbances in cell growth that ultimately lead to disease.

Proto-oncogene [☞ "Oncogene"]

What causes cancer at the molecular level? One answer lies in so-called proto-oncogenes, which are normal genes with normal functions in the cell; for example, regulation and control of cell division and cellular growth rates.

Proto-oncogenes gone bad, for example, through mutation or chromosomal translocation, become oncogenes. Oncogenes have lost their normal functions or have them constitutively (always) switched into the "on" position, not unlike a car with no brakes. The end result is the loss of cellular growth control, leading to the formation of different kinds of cancer.

Pseudogene

We retain vestiges of genes left in our genomes whose function has been selected against (deemed unnecessary) by evolution. Non-functional parts of these genes just keep hanging around in this long elimination process from generation to generation. These vestigial genes are called pseudogenes; they don't give rise to any functional gene products.

Pseudogenes can be a complicating nuisance when designing a diagnostic laboratory test searching for a particular gene or gene sequence of interest in a particular disease. It is possible that a pseudogene sequence may interfere with a reliable test result due to shared homology with the real gene.

Purines

The class of nucleotide bases that includes adenine and guanine; (see Figure 30) [☞ "Nucleotide"].

Pyrimidines

Another class of nucleotide bases that includes cytosine, thymine, and uracil (uracil is found naturally in RNA); because A base pairs with T (or U in RNA) and G with C, it follows that base pairing between strands of nucleic acid is between a purine and a pyrimidine (see Figure 30) [☞ "Nucleotide"].

Pyrosequencing

[☞ "DNA sequencing"]

Pyrimidines

- 6-member ring
- cytosine, thymine, uracil

Purines

- fused 5- and 6-member rings
- adenine, guanine

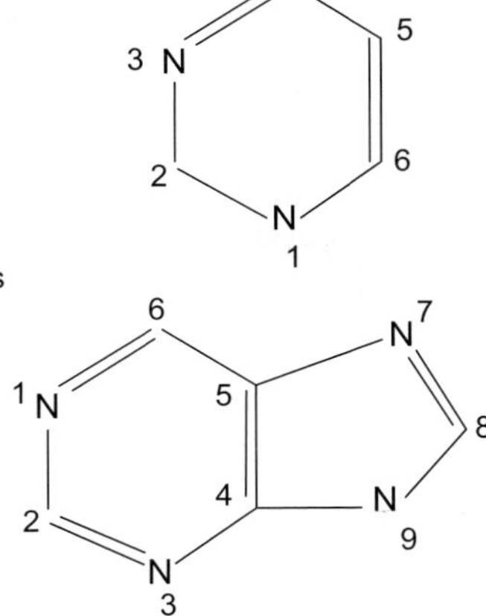

Figure 30. These nitrogenous bases are components of DNA structure.

Reprinted with permission from Tsongalis GT, Coleman WB. Molecular diagnostics: a training and study guide. Washington, DC: AACC Press, 2002.

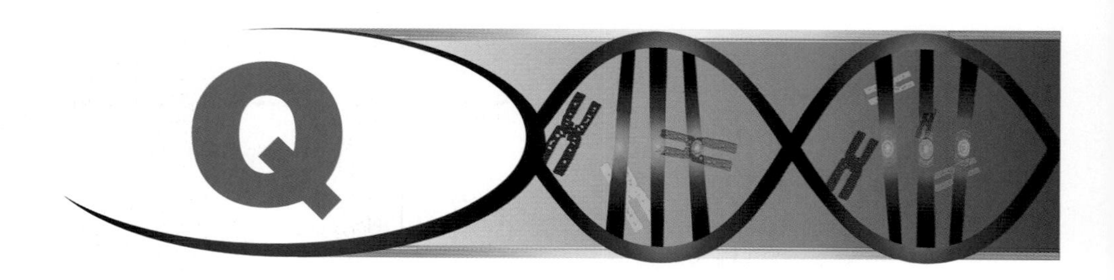

Qβ replicase

Qβ replicase is an enzyme; it is an RNA-dependent RNA polymerase, which means that it makes RNA daughter strands using a parent RNA strand as template. The enzyme is naturally found in a bacteriophage [☞ "Bacteriophage"] called Qβ. Commercial attempts to exploit this enzymatic system for a "PCR wannabe" [☞ "PCR wannabes"] have failed to date.

Quantitative PCR

In order to understand quantitative PCR, one must have an understanding of real-time PCR, which simultaneously accomplishes amplification of the DNA target and its detection as it occurs, i.e., in "real-time" [☞ "Real-time PCR"]. Quantitative PCR, or q-PCR, takes advantage of this ability to detect the amplified product *as soon as it happens*.

Detection occurs by measuring an increase in fluorescent signal once it reaches a level above that of background fluorescence. That point is known as the cycle threshold or C_T value, and its value is based on the amount of starting sample. If the sample contains a large amount of target, e.g., high viral load, a positive signal and crossing of the threshold occurs *sooner* than if the original sample had much *less* target. Because this is true, calibrators containing known amounts of starting template may be used to generate a standard curve, and this curve can be used to calculate the amount of starting material in the patient sample, thereby making the test quantitative [☞ "Cycle threshold (C_T) value;" "Real-time PCR"].

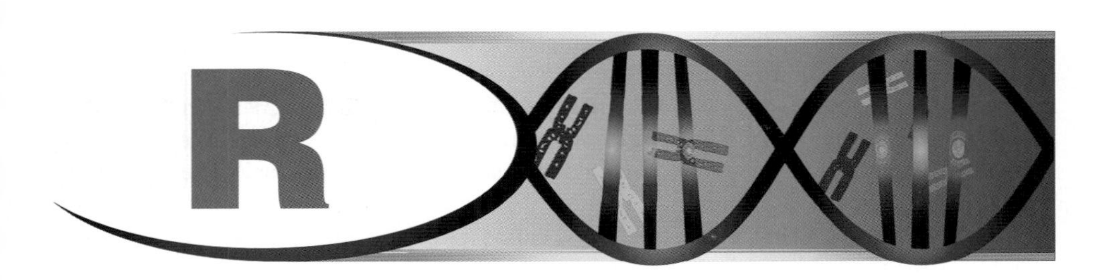

Random priming

Amplification of DNA or RNA usually requires knowledge of the nucleotide sequence of the target of interest. If the target sequence is not known, this can create a problem. Random priming allows amplification of unknown sequences using short oligonucleotide primers, usually six to eight base pairs in length, in the hopes that enough of a match will occur between the target and unknown template sequence to allow amplification.

Think of it as buying a lottery ticket, where six numbers are randomly chosen in an attempt to hit the big jackpot. If all six numbers aren't an exact match, there are still chances of winning by matching five numbers, or four numbers.

One of the most common methods for random priming uses random hexamers, which are every possible combination of the four nucleotides (A, C, T, or G) to give 4^6 possible primer combinations. Using these random hexamers, unknown sequences can be amplified and then "stitched" together to create a whole sequence. Scientists have discovered many new genes, which otherwise might have been difficult to isolate, using this method [☞ "PCR" and "Primer"].

Real-time PCR

Generally speaking, a successful molecular diagnostic laboratory test result depends on three things: (1) DNA extraction, (2) DNA amplification, and ultimately (3) DNA detection. Until real-time PCR became popular at the end of the last decade, these three phases of DNA diagnostics were discrete entities performed independently of each other. Real-time PCR successfully combines steps two and three. In other words, once DNA has been purified, it may be subjected to real-time PCR in an instrument (available from many different vendors) that both amplifies the DNA target and simultaneously detects those targets as they are generated cycle by cycle, i.e., in real time.

How is this accomplished? The answer lies in a new component of real-time PCR not present in conventional PCR, namely the reporter molecule. In real-time PCR, the reporter is detected by fluorescence, and these detectors, called fluorimeters, have been successfully engineered into the many instruments in the marketplace. The combination of fluorescent reporter molecules in the PCR mix with on-board fluorescence detection is at the heart of real-time PCR.

If the reporter molecule used in real-time PCR is fluorescent, how is it exploited to detect specific amplification? In other words, why isn't fluorescence nonspecifically detected regardless of the presence or absence of amplicon? There are a number of answers to this question and they differ based on the scheme being used.

The first scheme of detection is the simplest and depends on specific DNA dyes like SYBR Green or ethidium bromide. These dyes intercalate (insert) themselves into the stacked bases of double-stranded (ds) DNA and their natural fluorescence increases when inserted into dsDNA. As PCR generates more and more dsDNA each cycle, more SYBR Green (for example) becomes bound, and that increase in fluorescence is detected and quantified by the on-board fluorimeter.

Another popular scheme of detection is called TaqMan® and is in fact similar to the popular '80s video game called PacMan, but of course it makes use of *Taq* polymerase, hence the name. At the annealing temperature in real-time PCR, not only do the primers anneal but so does another component called a TaqMan probe. This probe is a short oligonucleotide probe labeled with reporter and quencher dyes designed to anneal to the target sequence downstream from one of the primers. Think of the DNA polymerase in PCR as a sphere, similar in appearance to the video game PacMan or Ms. PacMan character. As DNA polymerase moves along the template DNA strand synthesizing a new daughter strand, if it encounters a double-stranded area, which it in fact does when it encounters the TaqMan probe, the so-called 5' to 3' exonuclease activity of DNA polymerase kicks into action. The probe is "chewed" off the template strand as the new daughter strand is synthesized and the net result is that the fluorescent reporter dye is liberated from its nearby quencher so that fluorescence occurs, and is detected and quantified. The amount of fluorescence generated is detected in real time and is proportional to the amount of specific target, amplicon, generated during PCR.

Another popular variation is to use two probes that anneal to the area between the primers. The binding of both probes to the target sequence within two to four bases of each other is required for this reaction to occur. The probes have on their adjacent ends a "donor" and an "acceptor" molecule, instead of a reporter and a quencher molecule as described in the example above. When the two probes bind the target, the donor transfers energy to the acceptor, which excites it, resulting in fluorescence. Because of energy transfer, this method is often called two-probe FRET—for fluorescence resonance energy transfer. There are also other variations on the theme of real-time PCR [☞ "Beacons;" "Scorpions"].

Because the amount of fluorescence generated during real-time PCR is proportional to the amount of input target DNA, real-time PCR is quantitative. Some, therefore, refer to it as q-PCR to minimize the potential for confusion with the acronym, RT-PCR, which has traditionally meant reverse transcriptase PCR [☞ "Quantitative PCR"].

It is also possible to accomplish real-time PCR with another amplification technology: nucleic acid sequence based amplification (NASBA™, now sold under the label NucliSens by BioMerieux [☞ "NASBA™"]). Real-time PCR is among the most significant advances in molecular diagnostics in the early part of the 21st century [☞ "PCR"].

Recessive

[☞ "Inheritance"]

Recombinant DNA (rDNA; not to be confused with rRNA)

Recombinant DNA is the term used to describe artificially created DNA for purposes of genetic engineering. The goals of genetic engineering may be industrial scale production of therapeutic proteins, creation of medical laboratory reagents, research, and more [☞ "Genetic engineering;" "Restriction endonucleases"].

Recombinant DNA is sometimes abbreviated rDNA. That can be confusing, though, because rDNA is also the abbreviation for that portion of the genome that encodes the sequence of ribosomal RNA (a constituent of ribosomes, which are part of the protein-synthesizing machinery of the cell). Ribosomal RNA is abbreviated rRNA.

Replication

The Trekkies among you know that replication is the process of making food and drink in those fancy holes in the wall of the Enterprise on Star Trek. With respect to molecular biology however, replication is the process of making more DNA. Replication, a natural process inside cells, has been harnessed as a tool in the diagnostic molecular pathology laboratory through the use of DNA and RNA polymerases [☞ "Polymerase"].

Research use only (RUO)

The road to a clinical diagnostic test that is fully-approved by the FDA begins in the research laboratory. Scientists select the target of analysis, determine the best material on which to perform the analysis, decide the best analytical method, and then review the resulting data for accuracy and precision. Part of this process includes clinical laboratory personnel testing real patient samples. Since the performance characteristics of the assay are being determined, the results cannot be used for patient care, thus the assay is designated as research use only (RUO) during this data-gathering period.

Restriction endonucleases

Abbreviated RE; also called restriction enzymes (see Figure 31).

Restriction endonucleases belong to the general class of enzymes known as nucleases [☞ "Nuclease"]. Hundreds of REs are produced commercially and made available to molecular

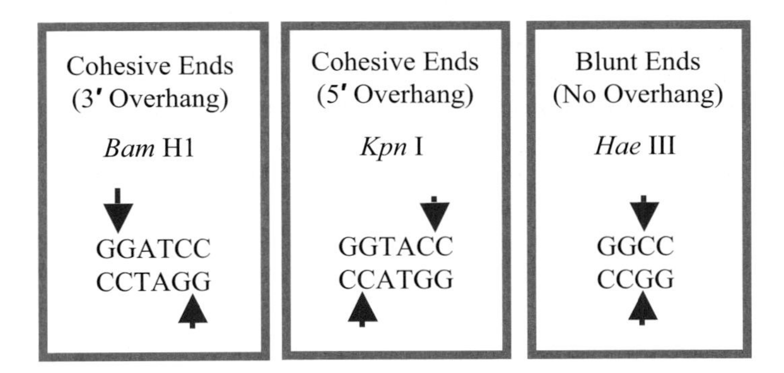

Figure 31. Examples of three different restriction enzymes, *Bam* HI, *Kpn* I, and *Hae* III. Arrows indicate exactly where in the DNA these enzymes cut, thereby generating the kinds of ends described.

Reprinted with permission from Tsongalis GT, Coleman WB. Molecular diagnostics: a training and study guide. Washington, DC: AACC Press, 2002.

pathology and molecular biology laboratories for use as tools in manipulating DNA. About 30 years ago REs had to be painstakingly and manually purified in order to proceed with the work of molecular biology. The current abundance and commercialization of REs is an important contributing factor to why clinical molecular pathology exists at all.

REs are naturally occurring bacterial proteins. Bacterial REs recognize unique stretches of bases, or sites (often, but not always, six base pairs long), in foreign DNA, and they cleave DNA at or near these sites. Some REs are sensitive to methylation patterns in their recognition sequences (a methyl group is a block of one carbon and three hydrogens attached to the DNA).

REs are the bacterial immune system. When a bacterium is infected by a bacterial virus (bacteriophage), one way in which that infection can be defeated is for the bacterial RE to recognize the viral DNA as foreign, and cut and inactivate it, thereby preventing the virus from doing its dirty work. Differences in methylation signal the RE to cut viral DNA and not the bacterium's own DNA. We have learned how to make use of REs for medical diagnostics and genetic engineering.

Recombinant DNA technology has depended on REs because when a piece of DNA is cut by such an enzyme it leaves an end (sometimes called a "sticky end") on the cut DNA molecule that fits very nicely and can be neatly "sewn" into any one of a number of different cloning vectors, like plasmids or YACs. These so-called recombinant DNA molecules can be mass produced.

REs digest DNA specifically, such that unique DNA restriction fragment families are generated by each enzyme. This in vitro biochemical reaction generates a range of DNA fragments differing in molecular weight and is the basis for the Southern blot technique that was so important in traditional clinical molecular pathology.

REs are named based on the bacteria from which they are purified. For example, the enzyme, *Eco* RI, is purified from a particular strain of *Escherichia coli*.

Retroviruses

These are viruses with an RNA genome that is converted by an enzyme called reverse transcriptase into a DNA intermediate [☞ also "Reverse transcriptase"].

That DNA intermediate can become more or less permanently integrated into the host cell genome. During this part of the viral life cycle, the virus is known as a provirus. The virus can leave the proviral stage of its life cycle and synthesize-pirating the machinery of the cell to do its work-new viral proteins and nucleic acids for the creation of viral progeny, which then leave the infected cell and go on to infect new cells.

Retroviruses are RNA tumor viruses. They can cause tumors in the animals they infect, for example, monkeys, chickens, rats, and mice. Human T-cell lymphotropic virus, types 1 and 2 (HTLV-1 and HTLV-2), are human retroviruses associated with leukemia and lymphoma. HTLV-3 has had several names but the one most people associate with it is human immunodeficiency virus (HIV) and is the agent that causes acquired immunodeficiency syndrome, or AIDS.

Reverse transcriptase

Abbreviated RT, reverse transcriptase is a nucleic acid polymerase. In other words, it's an enzyme (protein) whose job is to make nucleic acid using another nucleic acid as a template for that synthesis. RT is an RNA-dependent DNA polymerase. It makes DNA using an RNA template to direct the synthesis of that DNA.

RT is the exception we wrote about in the entry "Expression." Since Francis Crick raised the notion of the central dogma of molecular biology, and for more than a decade after that, it was believed that the flow of genetic information proceeded in only one direction: DNA → RNA → Protein. That was eventually proven incorrect by observations published independently in 1970, by three scientists, two of whom would go on to win the Nobel Prize for this work. David Baltimore at the Massachusetts Institute of Technology and Howard Temin at the McArdle Laboratory for Cancer Research at the University of Wisconsin in Madison (working with his colleague Satoshi Mizutani) did their studies on a class of viruses known as RNA tumor viruses. These are viruses that have only RNA in their genomes (when examined outside the cells they infect) and that under proper conditions cause tumors in the animals they infect. Temin (with Mizutani) and Baltimore went on to show that these viruses contain an enzyme that can direct the synthesis of DNA using the original viral RNA as a template for that DNA synthesis. The enzyme, called RNA-dependent DNA polymerase, was nicknamed reverse transcriptase because it shattered the then-current dogma and showed that transcription could happen in a "reverse" way. Transcription became a more generalized term to indicate the formation of intermediary nucleic acid (usually RNA, but this showed that DNA could be that intermediate) that went on to direct the rest of the viral life cycle within the infected cell.

The flow of genetic information in this class of viruses, which went on to become known as "retroviruses," is RNA → DNA → RNA → Protein. Infecting viral RNA, through the action of virally encoded reverse transcriptase, is transcribed into a DNA intermediate that goes on to direct the synthesis of viral RNA (for progeny virus). Ultimately viral proteins for new viral progeny are also made using the host cell's protein-synthesizing machinery (that's what viruses do; they "pirate" the host's tools to do the bidding of the virus). Retroviruses may arrest in their life cycle at the DNA stage. The DNA becomes integrated into the host cell genome and is known at this point as a provirus [☞ "Expression;" "Retroviruses;" "Virus"].

RFLP testing

Restriction fragment length polymorphism (RFLP) testing; individuals are different in many ways including their DNA sequences.

Differences in DNA among individuals are called polymorphisms. The most basic polymorphism is the single nucleotide polymorphism [☞ "SNP"]. Another is the length polymorphism alluded to here. One individual may have a restriction endonuclease recognition site [☞ "Restriction endonucleases;" "Southern blot"] at a particular point in his or her DNA, while another person does not. The two sites in these two people are said to be polymorphic, or different, from each other. The difference may be detected by examining the sizes of the restriction fragments generated when those two individuals' DNAs are cut with the same restriction endonuclease and electrophoresed (a technique to separate the DNA fragments). One individual has the site and the enzyme cuts there, while the other individual's DNA, lacking the site, is not cut and the fragment is therefore longer; hence restriction fragment length polymorphism (difference).

RFLP analysis has applications in paternity testing, genetic disease, and routine molecular pathology investigation. RFLPs may be detected by Southern blot analysis or polymerase chain reaction (see Figure 32).

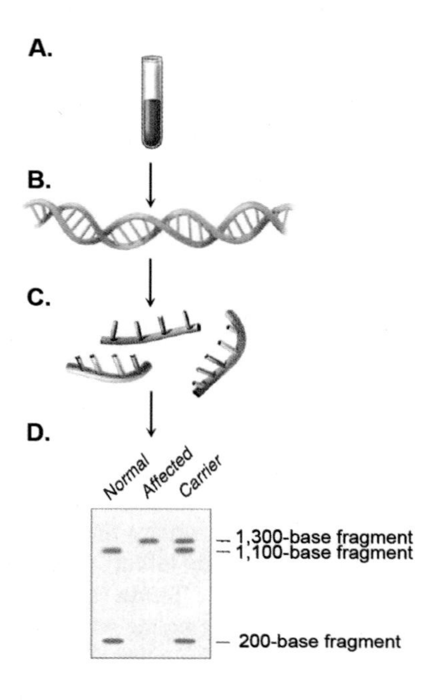

Figure 32. RFLP experiment. (A) Blood samples are taken, (B) DNA is isolated, and (C) cut into pieces with a restriction enzyme. (D) The resulting DNA fragments are then separated via electrophoresis. In this example, an *Mst*II-digested DNA sample was probed with the beta-globin gene, revealing the three "morphs" or forms possible for sickle cell anemia. The 1,300-base pair fragment indicates both the loss of an *Mst*II site and the presence of the mutation.

Ribonuclease (RNase)

[☞ "Nuclease"]

Ribosomal RNA (rRNA)

Ribosomal RNA, abbreviated rRNA, is a normal constituent of ribosomes.

Ribosomes are a significant component of the cell's protein-synthesizing machinery. Ribosomal RNA is among that class of RNA that has catalytic activity and it is the rRNA, not the protein, component of ribosomes that is responsible for ribosomal function [☞ "Ribosome"].

Ribosome

The ribosome is responsible for translating messenger RNA (mRNA) into protein.

This is accomplished by reading the genetic code that is embedded in the mRNA sequence in groups of three nucleotides and adding the correct amino acids in the correct order [☞ "Translation"]. The ribosome itself is made up of proteins and RNA, called rRNA (for ribosome). It is the rRNA, not the proteins, that is the functioning or catalytic part of the molecule.

Normally, ribosomes are very accurate macromolecular machines. Some messages however, contain strings of the same nucleotide, which can cause the message to become "slippery," causing ribosome shifting along the message. This shifting changes the reading frame and subsequent incorporation of different, incorrect amino acids.

Ribosomal frameshifting may be encoded in the mRNA transcripts, creating an intentional shift within the message, or may be unintentional. Intentional frameshifts are designed to get the most "bang for the buck" out of a single mRNA transcript and are common in RNA viruses with small genomes; something of an information packing scheme. Unintentional frameshifts, while not true genetic mutations, still result in an aberrant protein. Fortunately, most aberrant proteins are quickly identified and destroyed by cellular mechanisms. Human disease associated with ribosomal frameshifting is uncommon; however, some neurodegenerative diseases, including Alzheimer's, may be due in part to this mechanism.

Risk analysis

For decades, scientists have been trying to determine the genetic basis for disease. Even prior to the completion of the sequencing of the human genome in 2003, several diseases, cancers, and conditions were known to be associated with changes in DNA sequence. While scientists have made progress in understanding the genetic basis of many diseases, many more have yet to be elucidated. The fact that most diseases or conditions are not caused by a single, obvious mutation toughens the search.

One example is heart disease. What we know is that heart disease seems to run in families; however, specific mutations in one or more responsible genes have not yet been found. While some genetic associations have been identified in individuals having heart disease, other individuals with the same genetic sequence do not develop heart disease. We also know that in addition to genetics, other factors such as diet, exercise, and smoking habits play a role in whether or not a person will develop heart disease. Because of this, the physician is only able to tell patients their risk of getting heart disease based on their associated genetic mutation(s) (if any), family history, and lifestyle. Therefore, a risk analysis or risk assessment uses genetic information in conjunction with other factors in the hopes that patients will modify their lifestyles accordingly to reduce their risk of developing a specific disease or condition.

RNA

Ribonucleic acid; RNA is the transcriptional product of DNA and the intermediary for the flow of genetic information. RNA is also the genetic material of some viruses.

There are three main classes of RNA within cells:

1. Messenger RNA (mRNA) contains the blueprint for protein synthesis; that blueprint is read during translation.
2. Ribosomal RNA (rRNA) is a component of the machinery of translation, the ribosome.
3. Transfer RNA (tRNA) accomplishes the decoding of mRNA and adds an amino acid to the growing polypeptide (protein) chain during translation; it can be considered an adapter molecule that mediates the translation of RNA into protein.

Heterogeneous nuclear RNA (hnRNA) is another class that is composed of multiple sizes (hence the term heterogeneous), confined to the nucleus, and possessed of a short half-life. In other words, hnRNA is short-lived and includes among other things, mRNA precursors [☞ "DNA;" "Nucleic acids;" "Nucleotide"].

RNAi

RNAi stands for RNA interference and is a hot topic in the molecular world.

If one could selectively interfere with the expression of a specific RNA, or "knock down" its expression, one could look at the changes induced in the experimental model being used and learn about that gene transcript's role in the process being studied. For example, if one suspected that knocking down the expression of a particular gene caused cells in culture, which would otherwise die after a few rounds of cell division, to become immortal, one would learn a lot about the carcinogenic potential of that particular gene. This is the idea behind RNAi.

The principal biochemical tools used to achieve RNAi, and thereby learn about the role of a gene, are specific double-stranded RNAs (dsRNA) with complementarity to the gene of interest. Once introduced into a cell or organism (this has been done experimentally in worms, fruit flies, and plants), dsRNA induces the formation of small interfering RNA (siRNA). siRNA molecules are made through a series of biochemical reactions involving things with cool names like DICER (an enzyme that dices up the dsRNA into pieces that are then used to accomplish the interference) and RISC (RNA-inducing silencing complexes, which contain a ribonuclease sometimes referred to as SLICER). In this way, siRNAs allow important experimental questions to be asked and, we hope, answered. One very exciting potential application of RNAi is treatment of HIV infection. Researchers at MIT and Harvard Medical School have shown that RNAi can be used to suppress replication of human immunodeficiency virus in vitro. Perhaps some form of RNAi will be used in the future to control and suppress viral infections in patients.

Another type of RNA called microRNA or miRNA works in a similar manner to siRNA. miRNAs are encoded by the host DNA and transcribed into precursors, which then mature into miRNA. The miRNA is then cleaved by DICER, with the resulting products targeting host mRNA instead of exogeneous RNA. One difference between siRNA and miRNA is that the former is specific to a particular RNA, while the latter is non-specific and can be used to silence a variety of mRNAs.

An analogy to RNAi would be if one could selectively remove from a song the key track like the voice or key instrument. Imagine what "Stairway to Heaven" would sound like without the guitar track or what "White Christmas" would sound like with Bing Crosby's voice "knocked down" or suppressed. You would learn pretty quickly that the suppressed portion is absolutely vital to the proper expression of the song. The digital technique used to mask the key track is equivalent to siRNA and the result of this bastardized song is equivalent to RNAi.

RNase

[☞ "Nuclease"]

RT-PCR

Think of it as RNA-PCR. RT-PCR has conventionally come to mean reverse transcriptase-PCR and it is unfortunate that an equally popular molecular pathology laboratory technique, real-time PCR, can sometimes have the same acronym. Some use the acronym q-PCR to refer to the latter and reflect the characteristic that real-time PCR is quantitative [☞ "PCR"].

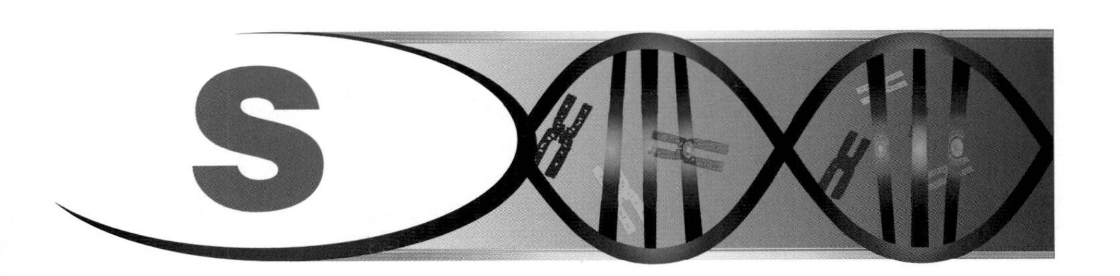

Safety

When handling patient specimens in the laboratory, we are always concerned about safety. In fact, we treat every specimen as if it actually is contaminated with human pathogens, like HIV-1 or hepatitis C virus, for example. For this reason human tissues, all of which *may* contain unknown infectious agents, are handled using "Universal Precautions." In 1996 the CDC modified these guidelines slightly and began referring to them as "Standard Precautions." When followed properly, these precautions minimize health risks associated with the handling of human specimens.

SARS

Severe acute respiratory syndrome; this viral disease causes a potentially fatal pneumonia and became newsworthy as it spread throughout the world in late 2002 and into 2003.

The causative agent is a single-stranded RNA virus called coronavirus (SARS-CoV). Since molecular diagnostics has matured to the point where we can quickly generate diagnostic assays based on the nucleic acid sequence of an organism, the diagnostics community was able to develop a SARS diagnostic test within weeks of the outbreak. Public health measures that stressed quarantine helped to minimize the spread of this pathogen. Fortunately, at the time of this writing in spring 2008, no additional SARS outbreaks have occurred.

This spread is an example of how, from a species point of view, *Homo sapiens* is very much at the mercy of nature's ability to adapt. Ultimately a pathogen or series of pathogens will arise, perhaps in the context of some cataclysmic natural or man-made event, against which we have no natural immunity and don't have or can't generate effective therapy. Bubonic plague wiped out tens of millions in medieval Europe. Influenza killed tens of millions in war-ravaged Europe and the rest of the world early in the twentieth century. HIV has killed more than 20 million since the current pandemic began in the early 1980s. Will avian flu or some other pathogen be next? Yes, probably; it's just a question of time.

One of the chief culprits in the potential for disaster is the overuse of antibiotics in developed societies. Antibiotics don't kill viruses and should not be prescribed or demanded for viral infections. Bacteria replicate very quickly and in mind-boggling numbers. During this propagation, mutations arise that confer resistance to our antibiotics. If we're not proactive, in the end the bacteria will win, because as a Klingon in Star Trek might say, what doesn't kill them makes them stronger. You can learn more at the FDA's website (www.fda.gov) by searching for "antibiotic resistance."

Scorpions

Scorpion primers/probes are named after the eight-legged arthropods that characteristically curl their tails over their bodies prior to stinging their victims. Scorpion primers/probes are molecules that contain a primer with a beacon-like probe attached to their 5'-end.

The primer acts like any other primer and initiates replication of the target sequence during the amplification process. Because the probe is attached, once the target sequence has been generated, like the arthropod, the probe part of the molecule "curls" over the primer and binds to the newly made sequence. Because of the close proximity of the probe to the sequence, the sensitivity and specificity of the assay are increased over other methods [☞ "Probe;" "Primer;" and "Beacons"].

Semiconservative

When you're young and your mind is full of utopian ideas, it's easy to have a liberal point of view on things. As one gets older and accumulates more, conservative thinking starts to creep into the psyche. Your authors are in that awkward transitional period—you could say we're semiconservative.

DNA is also semiconservative, at least with respect to its replication. During DNA replication, one strand of the double helix serves as the template for the synthesis of a new daughter strand. After replication is completed, there is one "old" strand of DNA that served as the template (and is conserved) and the complementary, new daughter strand: one old and one new. That's the thought behind terming DNA replication semiconservative.

The semiconservative nature of DNA replication was elucidated by Matthew Meselson and Frank Stahl working at the California Institute of Technology in Pasadena. They performed these experiments, now considered classic, in 1958.

Sense (and antisense)

Sense and antisense refer to the organization of bases in a DNA molecule (for example, a human gene) or an RNA molecule (for example, a viral gene). The sense strand is the one that is identical to the messenger RNA (mRNA) molecule that is synthesized through transcription and ultimately translated to a protein. Because the sense strand is identical to the mRNA (except that U replaces T in mRNA), it may be helpful to think of it as the coding strand. The antisense strand is the one complementary to the sense strand and is used as the template to direct the synthesis of the mRNA. Make sense? If not, see Figure 33.

Serum

When blood is drawn in a collection tube and there is no additive to prevent the blood from clotting, the liquid portion left after the clot forms is called serum [☞ "Plasma"].

Short tandem repeats

[☞ "Paternity/profiling/identity/forensic testing by DNA"]

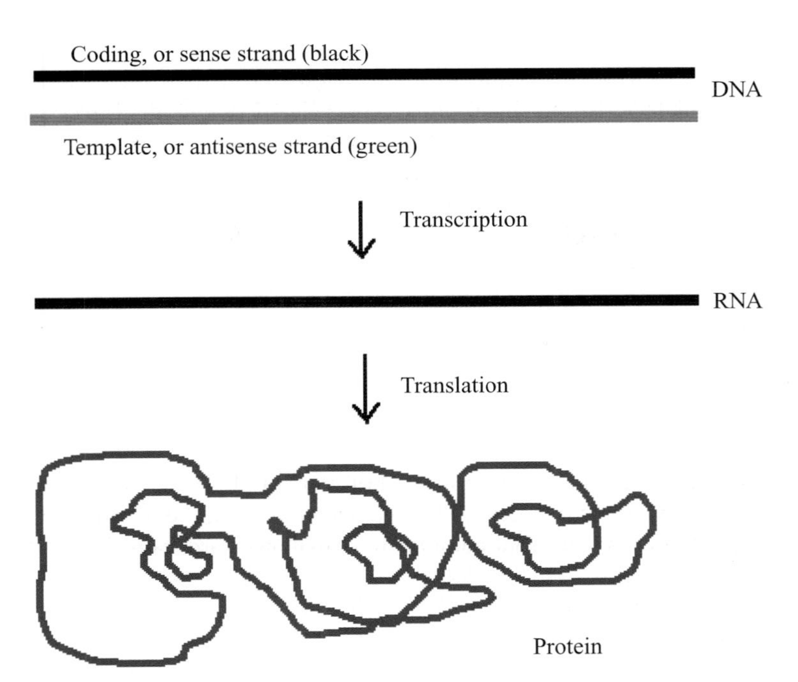

Figure 33. The coding DNA strand is indicated; it is the one that ultimately gives rise to the RNA that is translated into the gene product (almost always protein) of that gene.

Signal amplification

Two types of amplification schemes are widely used to analyze nucleic acids: those that amplify the target and those that amplify the signal. Target amplification techniques make more of the starting material and include methods such as PCR, transcription-mediated amplification (TMA), and nucleic acid sequenced-based amplification (NASBA). Signal amplification methods accentuate the target to make it more visible, but do not make more of the starting material. Some have likened signal amplification methods to putting lights on a Christmas tree—where the tree is the nucleic acid and the lights are the signal. Signal amplification methods include branched DNA (bDNA) and Hybrid Capture. [☞ "PCR wannabes"]

SNP

Single nucleotide polymorphism; the lowest common denominator among genomes. The term is pronounced "snip."

As discussed in the entry "Paternity testing," polymorphisms refer to different forms of the same basic structure. At the DNA level, polymorphisms are evident in different ways and the single nucleotide polymorphism refers to a difference of one base between sequences of two individuals. If the sequence CTAC is present as CT**T**C (note that one base has changed) in at least one percent of a particular population, it is defined as a SNP.

SNPs are not mutations, merely naturally occurring, evolutionarily stable (therefore heritable) differences. They are useful as markers within the genome for tracking disease in humans, growth characteristics in corn, or the economically important gene associated with litter size in pigs, for example. Some SNPs are markers for disease or propensity to disease, for example the (any one of several) SNPs in the *HFE* gene on chromosome 6 that, when present, cause hereditary hemochromatosis, an iron storage disease (learn more at www.americanhs.org/). So some SNPs are also deleterious (point) mutations. Not all SNPs are mutations and not all mutations are SNPs; some mutations are caused by other changes in the genome involving more than a single base pair change.

SNPs are responsible for most of human genetic variation and are observed about every 100 to 300 base pairs. If you divide three billion base pairs by 300, you come up with about 10 million SNPs in the human genome. While 10 million is still a large number, it's potentially a lot more powerful to look for certain SNP patterns diagnostic of disease in individuals with, for example, prostate cancer or diabetes, than it is to look for mutations in a single gene. That's because any one gene may have minimal impact on the causes of these multigenic disorders, and looking for certain heritable SNP patterns may prove to be more informative [☞ "Risk analysis"]. Once that association is available at a research level, then clinical diagnostic tools can be developed and, just as importantly, the various responsible genes can be identified and therapies contemplated.

Somatic cells [☞ "Germline"]

Somatic cells comprise all the cells in the body except for the germline cells (the sperm and the egg, or ovum) and undifferentiated stem cells. Somatic cells are a result of the fusion of two germline cells, an ovum and sperm, and contain a full set of chromosomes (23 pairs). Since somatic cells are the result of fusion between the sperm and the egg, any mutations present in those cells are also present in the somatic cells as "inheritable" mutations.

Somatic cells, however, may also contain "non-inheritable" mutations that arise during cell division. Depending on the stage of development when the mutations occur, the mutations may be present in the whole organism, or only present in a specific cell line. When these mutations occur, they are incorporated into all subsequent cells that are derived from the affected cell. While many mutations do arise spontaneously during cell division, they may also arise due to external factors like ultraviolet radiation, cigarette smoking, or chemical mutagens. Fortunately the cell has DNA repair enzymes that identify and correct most mutations before they cause harm. Even though these repair mechanisms exist, somatic cell mutations may occasionally become incorporated permanently into the DNA [☞ "Mutation"].

Southern blot (dot blot; slot blot)

The technique known as the Southern blot was developed by Dr. Edwin Southern in the mid-1970s in England (see Figure 34). It is a laboratory method used to identify "a needle in a haystack." More correctly, the technique is used to identify a single piece of hay within the haystack. Here's how it works.

Purified DNA is subjected to fragmentation with restriction endonucleases [☞ "Restriction endonucleases"]. Those fragments are then fractionated (separated) using gel electrophoresis [☞ "Electrophoresis"]. The gene or DNA sequence of interest is still buried

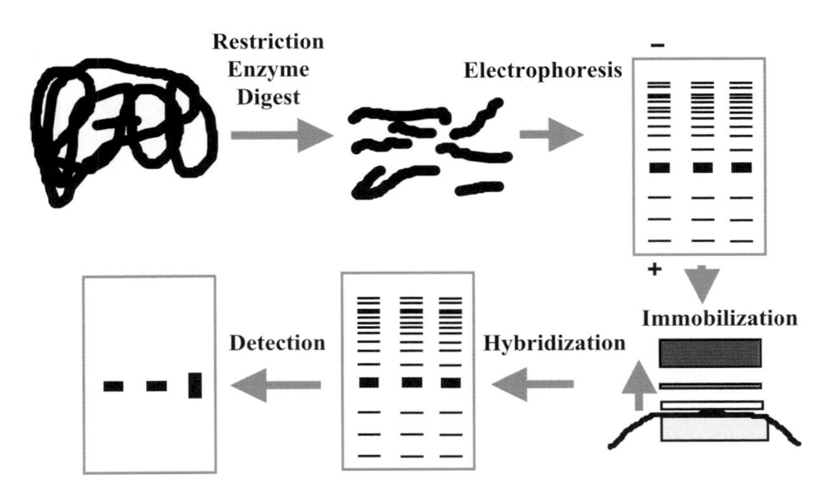

Figure 34. This cartoon depicts the steps in performing Southern blot analysis: DNA extraction followed by restriction enzyme digestion, electrophoretic fragment separation, transfer and immobilization, hybridization, and detection of the gene fragment of interest.

Reprinted with permission from Tsongalis GT, Coleman WB. Molecular diagnostics: a training and study guide. Washington, DC: AACC Press, 2002.

within that total genome's worth of fragmented, separated DNA. The next step is to transfer the DNA in the gel used for electrophoresis [☞ "Agarose"] to a more solid support. Early investigators used nitrocellulose paper but after a few years the field migrated toward the use of less fragile nylon membranes. Either way, a transfer of the DNA to a solid support that is little more than a very tough piece of special paper is accomplished by any one of a number of methods: vacuum (actually aspirating the DNA out of the gel and onto the paper); positive pressure; or capillary action (where salt water moves up from a reservoir through the DNA-containing gel and carries or transfers the DNA to a piece of nylon paper on top of the gel; the DNA binds to the paper).

After transfer is complete, the association between DNA and paper is tenuous but is made permanent by baking the Southern blot in an oven for an hour or two (about 80 °C). Once permanently attached, the DNA on the blot may be hybridized with a DNA probe to actually locate the DNA sequence of interest, the needle in the haystack [☞ "Autoradiograph;" "Probe"].

Southern blot-based testing is becoming obsolete in the clinical molecular pathology laboratory because of its labor intensity, long turnaround time (it can take many days to generate a result), and the poor reimbursement by insurance companies to clinical laboratories that perform this technique. Furthermore, PCR-based tests that answer most of the questions that were formerly answered by Southern blot investigation have been developed. There is still limited utility for Southern blot testing in B- and T-cell gene rearrangement analysis for leukemia and lymphoma diagnostics and fragile X syndrome analysis.

Dot and slot blots are variations on the theme of the Southern blot. They are the same except no restriction endonuclease digestion of DNA (and subsequent electrophoresis) is performed. Dot and slot blots, therefore, are used more for simple "yes or no" answers to this question: "Is a particular DNA sequence (for example, mutation or bacterial DNA) present in this sample?" Reverse dot blots have the probe affixed to spots on the nylon membrane. Amplified DNA is then added to the membrane and hybridization is detected by visualization of the reporter molecule built into the system. Depending on how the probe is affixed to the membrane, it may be called reverse dot blot or reverse line blot.

Note that the Southern blot is capitalized because it is named after a person. The "opposite" of the Southern blot examines RNA and is called the northern blot but is not capitalized because there was no "Dr. Northern." A related technique interrogates proteins and has therefore been given a related name, the western blot.

Splicing

One of us (DHF) had a brother, Steve, who used to love to play with his reel-to-reel tape recorder back in the '60s. He'd spend hours cutting and mending audio tape, and splicing it to create new recordings. Actually, that is the perfect analogy for how the term splicing is used with respect to DNA and RNA. The DNA in a mammalian, human, or plant gene is longer than the messenger RNA (mRNA) molecule derived from that gene. That's because of the presence of introns and exons [☞ "Exon;" "Intron"]. The introns are spliced out to form the mature mRNA sequence. The site between exon and intron in a gene is known as the splice junction. Splicing at the splice junction must be precise to the base because an error of even one base can disturb the reading frame of the resultant mRNA such that a mutant protein is produced. In fact, splice-site mutations are a class of mutations (see Figure 35).

SSSR (also known as 3SR)

Self sustained sequence replication is an unpopular in vitro nucleic acid amplification technique; it is a "PCR wannabe." The abbreviation is SSSR or 3SR.

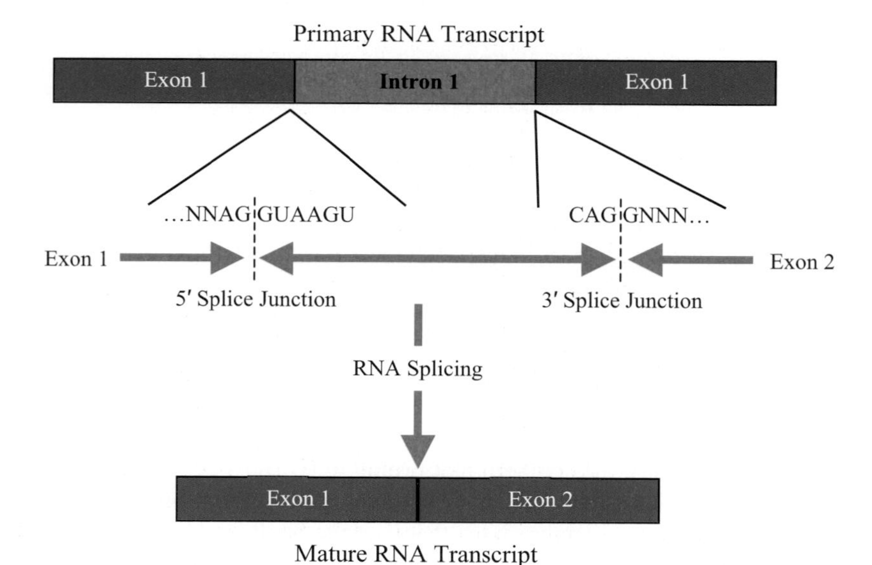

Figure 35. Fundamentals of RNA splicing. A= adenine; C = cytosine; G = guanine; N = any nucleotide; U = uracil.

Reprinted with permission from Tsongalis GT, Coleman WB. Molecular diagnostics: a training and study guide. Washington, DC: AACC Press, 2002.

Stem cells

Stem cells can be derived from adults or embryos and they have different roles in each. Stem cells are pluripotent; "pluri" means "many" and "potent" refers to "potential." Stem cells have multiple potential ways in which they can differentiate. That is, in embryos, made up of very few cells, of course, there has to be the potential to develop a heart, a pancreas, a brain, and all the different cell lineages in the blood and other systems. In other words, the concept of "unspecialization," the capacity to differentiate into any one of a number of cell types, is a general property of all stem cells. Another characteristic common to all stem cells is their capacity to divide and renew themselves over long periods of time. Stem cells may also be found in some adult tissues, for example, brain, muscle, and bone marrow. In these tissues, discrete stem cell populations exist to generate replacements for cells otherwise lost to normal wear and tear, injury, or disease. See Figure 36.

The pluripotentiality (now that's a word!) of stem cells makes them very exciting targets of research for therapy to replace diseased cells, for example to replace the diseased dopamine-producing neurons in the neurodegenerative disorder, Parkinson' disease.

At the same time, the very same characteristics so vital to the role of stem cells are the ones that could make them likely to seed cancers. In other words, stem cells can renew themselves over long periods of time; they are long-lived. During that time they could acquire the

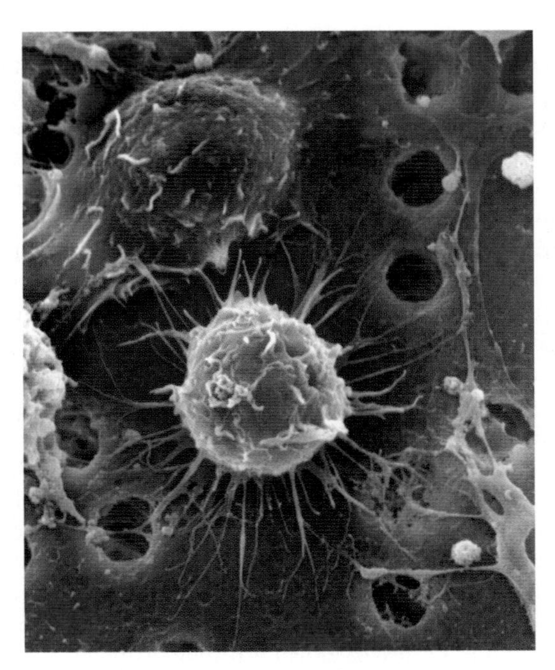

Figure 36. Human embryonic stem cell. Stem cells are the primal cells common to all multi-cellular organisms with the ability to renew themselves through cell division and can differentiate into specialized cell types. As a result of their ability to be transformed into specialized tissues such as muscles or nerves through cell culture, their use in medical therapies has been proposed with some controversy.

kinds of mutations that lead to cancer. In fact, mutant "cancer stem cells" have been found in human tumors. Because of this, we should apply caution in considering stem cells as therapeutic agents. At the same time, if cancer stem cells turn out to be a general phenomenon in all cancers, more rational thinking about cancer and new therapies may be the result.

This "yin and yang" of stem cell characteristics is a great example of the nature of science. The constant search for the truths of nature is an incredibly important quest if we are to harness nature's secrets to better humankind. Throughout the 1990s and early 21st century, the popular thinking was that stem cells, if properly harvested, could be a great boon to the treatment of disease; this may well be the case, in time. A spate of recent finding in the early 21st century, however, has sounded a note of caution. (One of the problems with books is that they quickly become out of date when reporting on ongoing scientific investigation. Those who pick up this book years after its 2008 publication will be at a disadvantage if they consider this piece on stem cells the end of the story. One might consider starting with the excellent summary by Jean Marx in *Science* magazine: "Mutant Stem Cells May Seed Cancer," volume 301, pages 1308–1310, September 5, 2003. The story continues from there, including the February 21, 2006, article in the New York Times: "Stem Cells May Be Key to Cancer.")

Synthetic genomics

Presumably, readers of this book consider genomics from a biomedical point of view. We would be negligent in our responsibilities as authors and teachers, however, if we did not point out one of the most exciting potential applications of genomics, which extends outside the biomedical arena and into the field of biologically produced fuels and energy. Bacteria, such as some found in the upper open ocean, metabolize carbon naturally. Others, like those found naturally in soil, generate hydrogen. Portions of the genomes of these bacteria are dedicated to achieving these biochemical processes. The technologies associated with the fields of molecular biology and genomics have matured to the point where scientists at the J. Craig Venter Institute (JCVI; www.jcvi.org) can now consider construction of artificial bacterial cells with just enough genome to sustain life, i.e., reproduce, and yet also produce biological products and renewable biofuels.

So the genomic adaptation of these artificial life forms would be the incorporation of the genetic capability (genotype) to metabolize vast quantities of carbon and/or generate vast quantities of hydrogen (desirable phenotypes). In other words, the relevant genes from, for example, the bacteria described above would be incorporated into these as yet (in 2008) hypothetical artificial organisms. (Rapid progress toward this end is being made; in early 2008, scientists at JCVI announced the construction of a common bacterial genome made up of nearly 600,000 base pairs. This genome is part of the path toward a fully synthetic organism that could then be manipulated as described.)

One application could involve elimination of carbon buildup in our environment, which has a negative impact and is a cause of global warming. A living "tool" that could be used to help minimize carbon buildup on this planet would have obvious benefits. Similarly, a living "tool" that could "naturally" generate a safe, environmentally friendly fuel source like hydrogen would also have far-reaching environmental and economic advantages (although the automobile industry would have to "retool" and the petroleum companies might have a word to say on the matter). Look for more and more "outside the box" applications of genomics in the future, particularly from Synthetic Genomics, Inc.

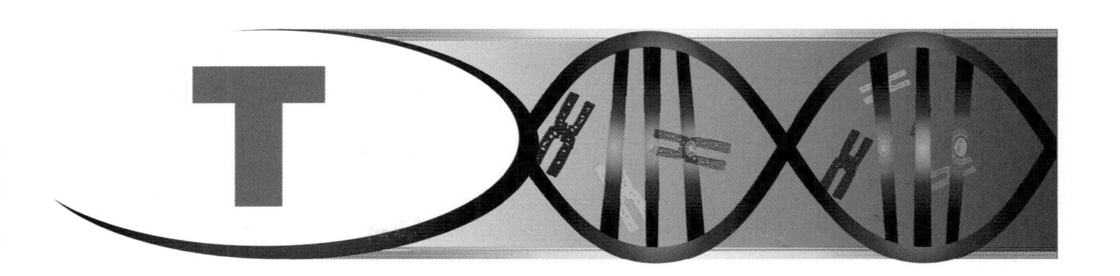

TaqMan primers

[☞ "Real-time PCR"]

Target

To a sharpshooter, the goal is hitting the ultimate target or "bullseye" in the center of a circle. Hitting the target in molecular biology refers to finding, using molecular tools described throughout this book, the correct part of the nucleic acid sequence being studied or analyzed.

The target is either DNA or RNA and may come from several sources, such as the host (of suspected disease); the (alleged) perpetrator (of a crime); or an infectious agent such as a virus or bacterium. For most nucleic acid sequences, the target sequence is a very, very small part of the whole genome.

Target amplification

Methods such as PCR create more of the starting product and are therefore called target amplification methods. Other commonly used types of target amplification methods include transcription-mediated amplification (TMA) and nucleic acid sequencebased amplification (NASBA) [☞ "PCR;" "PCR wannabes;" "NASBA"].

Telomere (and Telomerase)

It has been said that the only sure things are death and taxes. For the former, there is at least some molecular explanation for what happens to our cells. There is a limit to the number of times our cells can divide. As this limit is approached—in other words as our cells age—we age, ultimately leading to death. Thus mortality is defined at the DNA level.

The sequence TTAGGG is found at the ends of human chromosomes, the chromosomal region known as the telomere. TTAGGG repeats hundreds of times, effectively capping the chromosome end. As the cell ages, with each round of cell division some of the TTAGGG telomere units are lost, continually shortening the chromosomes. At some point late in life, so many telomeric units have been lost that the cell senesces, or goes into a resting stage.

Gametes, the reproductive cells (sperm and ova), do not succumb to this aging process and may be thought of as immortal. Though these cells can die, they do have the capacity to divide indefinitely. Cancer cells have the same capacity. One of the key factors in the immortality of these cells is an enzyme, telomerase (a specialized DNA polymerase), whose job is to replace the telomeric end units (TTAGGG) lost during cell division, thereby bestowing cellular immortality.

Using laboratory-engineered telomerase as a means to retard or reverse the aging process is not science fiction. Targeting telomerase activity in cancer cells to inhibit tumor growth and for use in other diseases is a prime area of investigation and potential commercialization by Geron Corporation (highly focused on stem cells) and its joint venture partner, TA Therapeutics, Ltd. Another angle on using telomerase in the clinic is to take advantage of assays to measure telomerase activity, either at the RNA or protein level. Remember, in most cells, there ought to be no telomerase protein or actively transcribing RNA that specifically codes for telomerase, even though the gene is present in all cells. Assays to detect the telomerase RNA or protein may prove effective in gauging a patient's response to various anti-tumor treatments. A decrease in telomerase would be a positive indication that therapy is effective.

Template

When DNA or RNA is replicated, either naturally in vivo, or artificially in the laboratory in vitro, new nucleic acid is being made, by definition. That synthesis is dependent upon the action of enzymes [☞ "Polymerase"]. The polymerase involved in the creation of new DNA or RNA is dependent upon on a master copy to direct the synthesis of the new strand of nucleic acid. That master copy is known as the template [☞ "Complementary strands of DNA"].

Thermal cycler

A thermal cycler is a microprocessor-controlled water bath that rapidly changes the temperatures among those needed to accomplish the polymerase chain reaction (PCR).

A typical PCR reaction may have to cycle among 94 °C, 55 °C, and 72 °C between 30 and 40 times. Thermal cyclers accomplish this cycling in an automated fashion. There are many fine commercially available models, and classic ones can even be purchased now at www.dovebid.com or on eBay™.

Thermus aquaticus

Dr. Thomas Brock and one of his undergraduate students, Hudson Freeze, first isolated the bacterium, *Thermus aquaticus*, from Mushroom Spring in Yellowstone National Park in the fall of 1966. Mushroom Spring is one of the park's hot springs—it has a temperature of 70–73 °C. The discovery of *T. aquaticus* and the subsequent isolation of its DNA polymerase was instrumental in the ability to automate the PCR process. This DNA polymerase, termed *Taq* polymerase after the bacterium (*T. aquaticus*), is the workhorse of PCR.

Before *Taq* polymerase was adapted for PCR, the technician performing PCR had to add new enzyme to the reaction at the start of each cycle because ordinary DNA polymerase, necessary for the reaction, could not withstand more than one cycle of high heat before denaturing and becoming inactive. This meant the technician had to closely attend the water baths (three different ones for each of the three temperatures), open the tube, add the enzyme, close the tube, transfer the tube among the water baths, and repeat this process 30 to 40 times. PCR was therefore a very labor intensive (and boring) process. In addition, manipulating the tubes often introduced contaminants.

The use of the *Taq* enzyme solved these problems. *Taq* polymerase is fully functional at temperatures over 90 °C, so it may be added at the start of PCR and it can be trusted to function for the entire experiment or assay, however many cycles are needed. Because of *Taq* polymerase, the entire PCR process could be automated; *Taq* is therefore a significant factor in the wide use of PCR in the clinical molecular pathology laboratory.

TIGR

TIGR stands for The Institute for Genomic Research. TIGR was a not-for-profit research institute in Rockville, MD, that was founded by Dr. J. Craig Venter and was involved in the structural, functional, and comparative analysis of genomes from viruses, bacteria, plants, animals, and humans. In 2006, TIGR merged with The Center for the Advancement of Genomics (TCAG), The J. Craig Venter Science Foundation, The Joint Technology Center, and the Institute for Biological Energy Alternatives to become the J. Craig Venter Institute (JCVI). Learn more about this outstanding research facility, which is doing groundbreaking genomics research, at its website (http://www.jcvi.org/). As we work on this entry in spring 2008, dozens of genomes have been sequenced by JCVI.

Tinman (if I only had a heart)

Basketball fans may remember the shocking deaths of Boston Celtics' star Reggie Lewis in mid-1993 and college player Hank Gathers in 1990. Both likely died of exercise-induced ventricular tachyarrhythmias (there is still controversy in the Lewis case); before the final diagnosis was made, however, hypertrophic cardiomyopathy (HCM) was also part of the differential diagnosis. In HCM, the heart's muscular wall thickens and though the heart still pumps blood strongly, the chamber-filling part of the heartbeat is adversely affected. Mutations in genes that encode myosin, a component of heart muscle, can cause HCM.

In mid-1998 the Seidmans and their molecular cardiology research group at Harvard Medical School described another mutation. Rather than myosin, this time the culprit was *TBX5*, a gene that encodes a transcription factor (a protein) involved in regulating other genes whose role is to build a healthy heart. The Harvard group found that *TBX5* mutations caused atrial-septal defects (holes in the heart). They also found that the human *TBX5* gene had significant homology (similarity) to a gene in the fruit fly (a commonly used animal model system to study genetics) called *tinman*. Fruit fly embryos that lack both copies of their *tinman* gene have no hearts at all (like the Tin Man in *The Wizard of Oz*); hence the creative name of the gene. Two copies of the *tinman* mutation, and presumably the *TBX5* mutation, are incompatible with life. Humans with one bad copy and one good copy of *TBX5* showed the hole-in-the-heart defects.

So although cigarette smoking, poor diet, sedentary lifestyles, etc., are enemies of good cardiovascular health, there is also a (potentially very strong) genetic component. The study of that genetic component will undoubtedly lead eventually to diagnostic and therapeutic approaches for those families [☞ "Proband"] in which hereditary cardiology problems exist.

And, at the same time, the study of these relatively rare heritable problems will serve as an excellent model for the wider problem of heart disease that affects so many.

You may notice that the title of this entry, *tinman*, is italicized because in this context, it is the name of a gene and gene names are italicized by convention.

The Wizard of Oz was selected by the American Film Institute as the sixth best film of all time when the list first came out in 1997; ten years later the film had dropped to tenth place (*Gone with the Wind* was the new number six).

Tissue-specific gene expression

Tissue-specific gene expression is tied to the differentiation of our cells from stem cells, thereby allowing the different cells in our bodies to form specialized tissue, organs, and organ systems.

All the cells in our body (except mature red blood cells and gametes; ☞ "Allele") contain the full measure of DNA that humans possess. Mature red blood cells do not contain DNA. Gametes (sperm and eggs) contain half the DNA found in a somatic cell, like a stomach, nerve, or liver cell. By "full measure" we mean that even though a stomach cell is specialized for the task of digesting food, it still has all the DNA that is involved in hair color, antibody production, vision, and everything else that our bodies do that depends on the expression of proteins encoded by DNA.

What makes a stomach cell a stomach cell and not a liver or nerve cell is that due to complex biology, endocrinology (the science of hormones), biochemistry, and other processes, tissue-specific gene expression occurs. A cell committed to being a stomach cell and finding itself in the biological environment of the stomach expresses only those genes in its full complement of DNA that are necessary for a stomach cell to develop and function properly. It leaves silent and unexpressed the rest of the DNA genome so that it doesn't start doing the job of a liver cell or a nerve cell, etc. In this example, the pluripotent stem cell from which the stomach cell was derived is said to have fully differentiated into a stomach cell [☞ "Stem cells"].

Tissue-specific gene expression is necessary to avoid total biological chaos.

T_m

T_m is the acronym used to denote the melting temperature of a DNA duplex. The "melting" refers to the dissociation of the two strands in the double helix.

T_m is defined mathematically as the temperature at which 50% of the double-stranded DNA molecules in solution are dissociated from each other and 50% are associated with each other. In other words, half the molecules are double-stranded and half are single-stranded. $G \equiv C$ base pairs in DNA are more stable than $A = T$ base pairs because G-C pairs have three hydrogen bonds holding them together and A-T pairs have only two. The higher the $G + C$ content of a particular piece of DNA, therefore, the more thermal energy (heat) required to dissociate the DNA strands and the higher the T_m.

T_m is used to differentiate mutations from normal sequences as well as pathogenic DNA targets in real-time PCR [☞ "Real-time PCR"]. Based on the sequence of the fluorescent reporter probes used in real-time PCR assays, fluorescence may be detected at different temperatures, during a so-called melting analysis, depending on the sequence of the target.

Perfect matches (wild-type probe hybridized to wild-type target) are more stable and have a higher T_m than imperfect matches (wild-type probe hybridized to mutant target). Perfect matches melt and therefore release their fluorescence in such assays at higher temperatures than imperfect ones. This temperature-specific fluorescence can be used to make an identification, i.e., a diagnosis.

Tobacco

Perhaps this entry should have gone under "Potatoes" or "Nutraceuticals," but we thought it might catch your eye to have a book significantly about health contain an entry for "Tobacco."

The word "nutraceuticals" is a hybrid between "nutrition" and "pharmaceutical" and meant to convey the notion of dietary supplements that fall somewhere between ordinary food-derived nutrition or dietary supplements like vitamins and the realm of pharmaceuticals. The potential of developing pharmaceutical agents from plants, a form of nutraceuticals, has been investigated. One study showed that transgenic [☞ "Transgene; Transgenic"] tobacco plants could be genetically engineered to produce an antibody to *Streptococcus mutans*, a bacterium that is a major contributor to tooth decay. The antibody was extracted from homogenized plants and inhibited re-colonization of the oral cavity with these bacteria in volunteers for four months. About one kilogram of tobacco was necessary for one complete treatment. That would be three kilograms of tobacco per year for every person who opted for this treatment at the local dentist's office, assuming the continuing research shows that this is viable. Indeed CaroRx™ is the derived drug that may (or may not) soon be on the market for this application.

Cigarette smoking thankfully continues to decline. At the same time, the resultant financial threat to U.S. tobacco farmers and the vested political interest of the relevant politicians from the tobacco states are realities. It seems there is great opportunity to turn a harmful product into a beneficial one while maintaining the economies associated with the tobacco industry. An interesting coalition of the dental community, tobacco farmers, politicians, and biotech and pharmaceutical companies could move this forward. Even the beleaguered cigarette companies could take advantage of an opportunity to shift from marketing a health hazard to something good for public health.

In other reports, potatoes have been engineered to express a subunit of the enterotoxin protein of enterotoxigenic *Escherichia coli*, a bacterium responsible for millions of deaths worldwide from severe diarrhea. Data from these studies suggested the feasibility of conferring immunity to this bacterium, and therefore to the disease, by eating a small bit of raw, appropriately transgenic potatoes.

For all of those reading this book who are concerned about eating transgenic plants, consider that every vegetable, fruit, meat product, etc., that one eats contains DNA. Transgenic potatoes, for example, have a bit of extra DNA in them, which is not ordinarily found in potatoes, but it's still DNA. It is not treated any differently by our digestive systems. The DNA, and resultant expressed protein, is broken down into its component parts in the mouth, stomach, and intestines. Continuing with this example, it is not inconceivable that some small fraction of individuals could be allergic to proteins not ordinarily found in potatoes, but such allergies are likely to be rare and could be assessed in advance to avoid allergic reactions.

Transgenic potatoes, tobacco, or sheep, for example, are simply organisms that were manipulated at the one-cell or few-cell stages, very early in embryogenesis, such that a foreign piece of DNA was inserted into their genomes. In this way the mature plant or animal has a bit of foreign DNA in every cell that is expressed through the normal gene expression mechanisms of that potato, tobacco plant, or sheep.

Toxicogenomics

In the same way that we can assess gene variations to determine how an individual will (or will not) respond to a drug (☞ "Pharmacogenomics"), so, too, can genes be interrogated to learn the likelihood of a toxic reaction. The study of how genes are mutated and how gene expression profiles change in cells exposed to chemicals or other sorts of toxins is a field rich with opportunity to learn more about the environmental hazards, for example.

Transcription

The synthesis of RNA, including messenger RNA (mRNA) from a DNA template, is the process known as transcription (see Figure 37). Just as a court recorder transcribes courtroom testimony so that the spoken word may be read, so, too, does the process of biochemical transcription turn DNA into RNA so that the DNA code may be "read" by the cell's protein-synthesizing machinery and ultimately translated into protein.

Interference with transcription is one mechanism by which cancer occurs. A human proto-oncogene called *myc* on chromosome 8 encodes a normal protein composed of 439 amino acids. The normal function of the myc protein is expressed during proliferation in

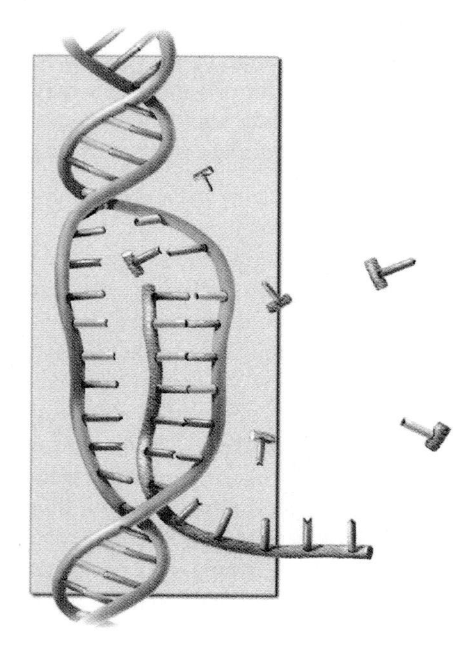

Figure 37. Transcription—the information in a gene is first converted, or transcribed, into the language of the messenger RNA (mRNA).

many adult tissues as well as during the development of the embryo. Examples of normal adult cell proliferation include regeneration of liver cells after hepatic injury or generation of new blood cells in the bone marrow to replace dying blood cells that simply "wear out." These are times when proliferation is a "good thing," just as in development of an embryo when such significant amounts of growth need to occur in such a short time.

Sometimes, as cells divide, chromosomal translocations occur. One example of such a translocation moves the *myc* proto-oncogene from its normal "address" on chromosome 8 to a location on chromosome 14 where it now comes under new transcriptional control. The cell has built-in checks and balances to regulate transcription. After an abnormal event like chromosomal translocation, ordinary regulation may be disturbed, as in the *myc* proto-oncogene example. In this case, *myc* proto-oncogene transcription is "up regulated" so that abnormally high amounts are made and for more sustained periods than normal. Because the normal function of the myc protein is to cause proliferation, it is clear that deregulated *myc* expression can lead to aberrant, increased proliferation, also known as cancer. In its new, abnormal address, the myc proto-oncogene becomes a cancer-causing oncogene [☞ "Chromosomal translocation;" "Oncogene;" "Proto-oncogene;" "Translation"].

Transcription-mediated amplification

TECHIE: Transcription-mediated amplification (TMA) is a "PCR wannabe" that has stood the test of time; it was developed at Gen-Probe in San Diego, CA. TMA may be used to amplify either RNA or DNA sequences using two primers and two enzymes.

The first primer is engineered by adding an RNA polymerase binding sequence to the 5' end of the primer. By doing this, the RNA polymerase binding site will become incorporated into the amplified product for use later in the reaction. This primer is complementary to the target rRNA on its 3' end and will direct synthesis of the first step. The first step in the TMA reaction uses this engineered primer along with the enzyme, reverse transcriptase, to create a cDNA strand. At this point in the assay, there is a rRNA:cDNA hybrid containing the newly created RNA polymerase binding site. In the next step, the rRNA template is removed via the RNase H activity of the reverse transcriptase enzyme, leaving only the cDNA molecule.

The following step exploits the second primer (this one does not contain any additional sequence) to create a cDNA:DNA molecule. Next, RNA polymerase binds and generates 100–1000 RNA copies—this is why the binding site was added and why it has transcription in the name. The newly made RNA copies then participate in a repetition of the process, generating more than a 10-billion fold amplification of the initial target in approximately one hour.

TMA has some advantages over PCR: it is isothermal (performed at one temperature, ~60 °C); is subject to less contamination; and it creates a thousand-fold more copies per reaction. One of the best commercial assays for *Chlamydia trachomatis* and *Neisseria gonorrhoeae* is based on this method [☞ "PCR wannabes;" "Transcription"].

Transgene; Transgenic

A transgenic animal is one that has had a foreign DNA sequence, i.e., a gene, introduced into it very early in the development of that animal, even as early as when the animal was a single cell. The process involves microinjecting DNA, under a microscope, into the cell that is held firmly in place.

The research opportunities afforded by transgenic animals have been bountiful, allowing us to learn more about gene expression, gene regulation, cancer formation, etc. The practical repercussions of all this may be very significant as commercial and research endeavors move toward developing medically useful transgenic animals. Examples include transgenic goats that have blood flowing through them with human proteins engineered in such a way that this blood is suitable for human blood transfusion. Transgenic pigs containing hearts with human surface proteins, which when transplanted into a human are not rejected by the patient as "foreign," are another example of a medically useful transgenic animal.

The medical implications are exciting. At the same time, as a society we need to deal with the issue of the ethical treatment of animals that will be raised to be donors of "humanized" organs (xenotransplantation). These are important questions that scientific progress is forcing us to consider.

Translation

The synthesis of proteins from an mRNA molecule using ribosomes and the rest of the cell's protein-synthesizing machinery is called translation (see Figure 38). The word "translation" is used because the genetic code, "written" in nucleotides, must be translated to the building blocks of proteins, which are "written" in amino acids.

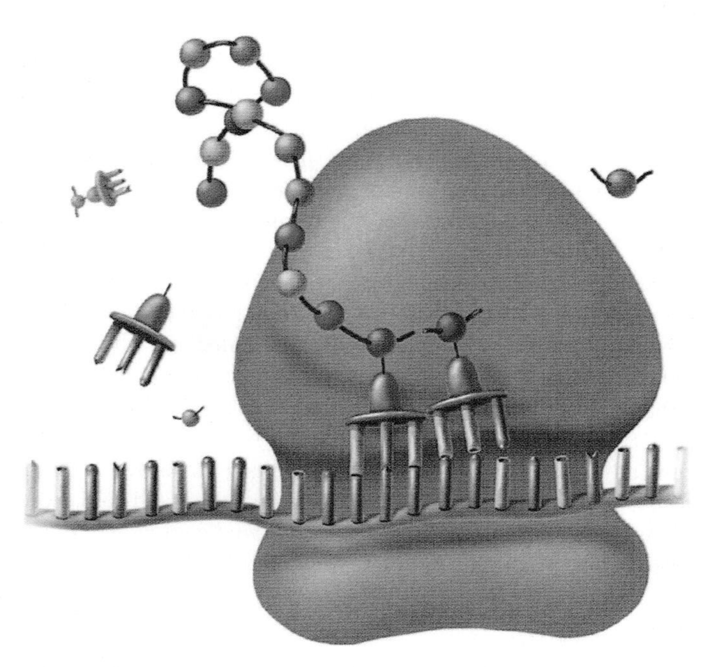

Figure 38. Translation—the ribosome is the large green cellular organelle in the center. As mRNA passes through it, the ribosome translates mRNA into a growing protein molecule (shown at the top of the ribosome).

Transposon

[☞ "Jumping genes"]

Trisomy [☞ "Monosomy"]

Normally, somatic cells have 23 pairs of chromosomes (one half of each pair from the mother, one half from the father). Occasionally, something goes wrong and instead of having two chromosomes, a person has an extra chromosome or part of a chromosome, such that one of the pairs now has three identical chromosomes. This is known as trisomy. Trisomy generally results from an abnormal gamete cell (sperm or egg) that contains the extra chromosome due to faulty chromosomal division during its production.

An extra, third chromosome is responsible for several medical syndromes. The most common is Down syndrome, where there is an extra chromosome 21: Trisomy 21. Other syndromes include Edwards (Trisomy 18); Patau (Trisomy 13); Warkany (Trisomy 8); Triple X (XXX, seen in females); and Klinefelter's (XXY, seen in males). Autosomal (non-sex chromosomes) trisomy syndromes generally result in distinct physical appearances (slanted eyes, short stature, wide noses, etc.); medical problems (congenital heart disease, thyroid disorders); and mental retardation.

Trisomy of the sex chromosomes (the X and Y chromosomes) generally do not have any associated symptoms. These individuals probably don't even know they are affected.

Tumor suppressor genes

Tumor suppressor genes play a role in the pathogenesis (defined as development of disease) of different cancers that occur as rarely as retinoblastoma (RB; an eye tumor) and as frequently as colon cancer.

Normally there are two alleles of a gene called *RB* on chromosome 13. Loss of one copy, through mutation, does not affect the individual or lead to cancer, but loss of the second allele, also through mutation, leads to deregulated cell growth and retinoblastoma. The normal function of *RB* is to suppress the growth of this kind of tumor. When that gene's function is lost due to mutation, the normal check on tumor development is lost. It's like losing the brakes on a car. On the other hand, mutations in proto-oncogenes (which make them cancer-causing oncogenes) are like having the accelerator pedal on a car stuck so that constant unchecked growth is the result [☞ the *myc* discussion under "Transcription"].

The *p53* tumor suppressor gene, when mutated, is involved in many different kinds of cancer, including breast cancer. A test for *p53* mutations is useful in diagnosing cancer risk so that early intervention, vigorous monitoring, and possible avoidance may occur. The *p53* gene is so named because the gene product is a protein that is 53,000 daltons in mass. The normal function of p53 protein is to suppress the multiplication of cells that have DNA damage until DNA proofreading mechanisms inside the cell repair the DNA or the cell commits suicide [☞ "Apoptosis"]. If p53 doesn't function, due to *p53* gene mutation, then other cancer-causing mutations that may be present in the cell are not stopped in their tracks and disease ensues.

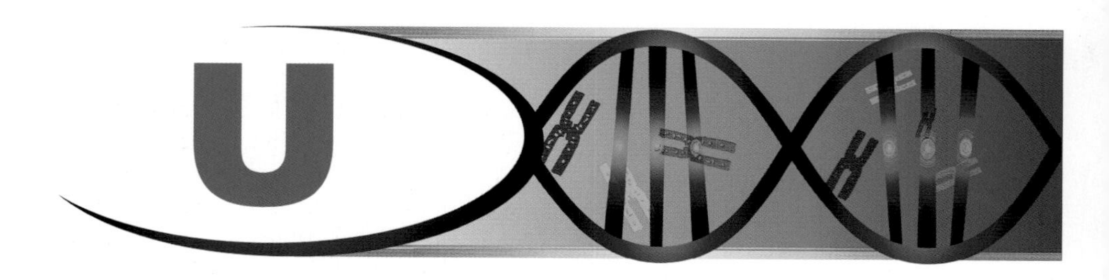

UNG

Uracil-DNA N-glycosylase (UNG) is an enzyme with the unique job of recognizing and destroying DNA molecules containing the base uracil.

"Why would DNA contain the base uracil?" you might ask. (In case you missed this part, uracil is naturally found in RNA, not DNA.) Well, normally it wouldn't; however, if one adds uracil (dUTP) to a PCR reaction in the laboratory instead of the usual base thymidine (dTTP), then the dUTP will be forced to base pair with adenine because there isn't any dTTP around. "Why would one do that?" you would ask next. Well, before we answer this question, we need to change topics a bit to discuss laboratory-based contamination of PCR reactions.

Contamination of negative specimens can be a nightmare for the professionals who work in a clinical laboratory, because it can cause false positive results. Contamination can occur during any of the many steps involved in the analysis of the sample—from other specimens, or from equipment, reagents, or previously amplified specimens. Since the purpose of an amplification reaction is to create millions of copies of the starting target, any manipulation of that target, such as opening the tube to perform a detection step, can create aerosols that can contaminate new specimens; specimens that would otherwise test negative for a particular analyte might test falsely positive if contaminated.

In an attempt to prevent contamination from amplicon carryover (the amplified target is called an amplicon), dUTP is added to the reaction in the place of dTTP. The resulting DNA made during the PCR amplification process incorporates the dUTP, creating a unique DNA amplicon. If this unique dUTP containing DNA amplicon contaminates a new specimen, it will be recognized by the enzyme UNG and be destroyed, thereby "sterilizing" the new reaction. UNG is added to the new reaction along with all the normal PCR reagents and incubated for a period of time long enough for the enzyme to destroy any dUTP-containing amplicons; since dUTP is not naturally present in DNA, only contaminating DNA from a previous amplification reaction containing dUTP will be recognized by the UNG, thereby sterilizing the reaction against contamination. After this initial incubation step, the reaction is heated, inactivating the UNG, but not the PCR reagents. After the UNG is inactivated, the normal PCR reaction takes place, again with dUTP in the mix to prevent carryover to the next set of specimens.

Contamination is much less of a concern now that real-time PCR dominates the clinical laboratory as opposed to conventional PCR that dominated in the '90s [☞ "PCR;" "Real-time PCR].

Untranslated region (UTR)

This is the part of an mRNA molecule that does not get translated into protein. Untranslated regions (UTRs) include the sequences that bind mRNA to the ribosome, orienting it for correct initiation of protein synthesis, and the "tail" that does not code for amino acids.

If the UTR is at the beginning of the mRNA it is called the 5'-UTR; if at the end, it is called the 3'-UTR. Mutations in the 5'-UTR can affect mRNA translation; some mutations "down-regulate" translation (less protein is made) while some "up-regulate" translation, resulting in more protein than would otherwise be made. Many diseases are caused by such translational 5'-UTR mutations and are in the new category of translational pathophysiology diseases. Diseases associated with translational mutations include Alzheimer's, cataracts, and some melanomas.

UTRs are also commonly found in viruses and are often highly conserved (have similar sequences) [☞ "Conserved"]. Because they are so conserved, they make good target sequences in diagnostic assays, especially for the RNA viruses, which tend to have a high amount of variation in the coding portions of their genomes. Conserved sequences are therefore targeted in many commercially available RNA viral assays such as those for hepatitis C virus (HCV).

Upstream

You don't need a PhD for this one. "Upstream" is the opposite of (all together now), yes, that's right—downstream. See the "Downstream" entry for an explanation.

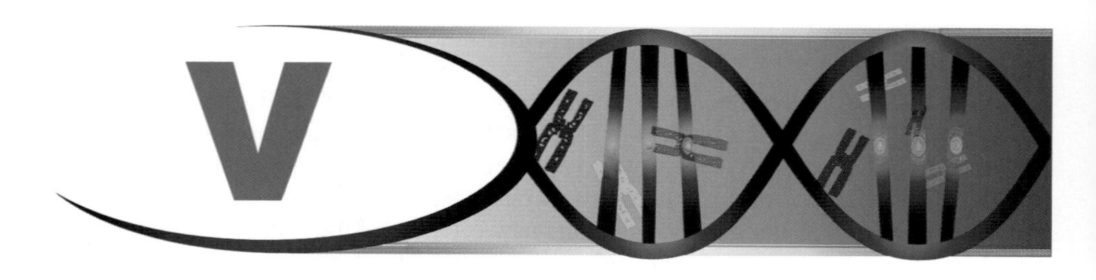

Variable number tandem repeats (VNTR)

[☞ "Paternity/profiling/identity/forensic testing by DNA"]

Vector

"Vector" is translated from the Latin as "bearer." In the case of molecular biology, a vector is usually a piece of DNA that accepts inserts of foreign DNA and carries them along with their own genome. Vectors include plasmids, cosmids, and bacterial/human/yeast artificial chromosomes (BAC, HAC and YACs) [☞ "BAC;" "HAC;" "YAC;" "Cosmid;" "Plasmid"].

Vectors are also agents that carry infectious diseases such as insects, arthropods, small animals, and yes, even humans. Examples of vector-borne diseases include malaria, yellow fever, tularemia, and Lyme disease. For a list of vector-borne diseases visit the CDC website (Division of Vector Borne Infectious Diseases) at http://www.cdc.gov/ncidod/dvbid/.

Viral genotyping [☞ "DNA sequencing"]

Viral genotyping as a diagnostic test in the clinical laboratory is applied primarily to Human Immunodeficiency Virus (HIV-1), the causative agent of AIDS. The concepts are similar to those described in the entry for "Antibiotic resistance" (of course, HIV is a virus and antibiotics are ineffectual).

Dozens of different drugs are used to combat HIV infection. These drugs kill most of the virus in the infected patient but at the same time, resistant forms of HIV will survive; they are said to be "selected for" by the drug. In other words, the killing pressure exerted by the drugs forces HIV to adapt by mutating its genome to give rise to different viral characteristics that make the virus resistant to the drug(s). The phenotype (characteristic) is drug resistance. The change in the genome, or genotype, is what gives rise to the phenotype.

There are a number of ways to inspect the viral genome for these characteristic mutations, but the very best way (the gold standard) is to check the sequence of the virus by a method known as DNA sequencing [☞ "DNA sequencing"]. Comparing the obtained results to a database of mutations known to correlate with resistance to different anti-HIV drugs offers the pathologist important information about what drug resistance is beginning to emerge in the HIV infecting the patient. The molecular pathologist, as medical consultant, conveys that information to the treating physician so that (s)he can more rationally deal with that patient's illness.

Increasingly, viral genotyping is being applied to hepatitis C virus (HCV). Using molecular diagnostic techniques, assessment of the HCV genotype is information a treating physician may use to determine the likelihood of success of interferon therapy to treat the infection and the appropriate duration of therapy.

Viral load testing

Using molecular technology (nucleic acid extraction followed by specific amplification of viral RNA), the clinical molecular pathology laboratory can measure the number of viral particles in a patient's blood specimen. Actually, blood is usually not the specimen of choice, but rather it is plasma or serum [☞ "Plasma;" "Serum"]. So-called viral load testing or viral load determination is a key element of managing HIV-1/AIDS.

An undetectable or constantly low viral load ensures the patient and treating physician that the infection is in check and that the multi-drug, antiviral cocktail being used for treatment is effective. An increase in viral load indicates (1) non-compliance on the part of the patient with respect to taking the medication or (2) viral drug resistance, meaning that the choices in the drug cocktail need to be re-evaluated [☞ "Viral genotyping"]. Patients being treated for HIV-1 infection have periodic viral load tests.

Viral load testing is also important in managing hepatitis C virus (HCV) and hepatitis B virus (HBV) infections. Viral load testing is performed with PCR-based tests available from Roche Molecular Systems, bDNA tests available from Siemens, or NASBA-based tests available from bioMerieux.

These viruses may also be detected by another in vitro nucleic acid amplification technique called Transcription-Mediated Amplification (Gen-Probe, San Diego, CA) in an instrument designed to test units of blood before they are placed in a blood bank.

Virus

Those who are inclined to ponder such things generally consider viruses to be the simplest and most basic form of life, although the prions of Mad Cow disease fame would, we think, argue. Viruses reproduce (one of the most basic definitions of life) by commandeering the biochemical machinery of the cells they infect. The viruses that pirate animal cells are called animal viruses (animal cells are categorized as eukaryotic cells that have a distinct nucleus and several other characteristic features). The viruses that infect bacterial cells are called bacterial viruses or bacteriophage (bacteria are classified as prokaryotic cells that are much more primitive than eukaryotic cells). Viruses cannot reproduce without first entering a host cell. They are therefore considered obligate intracellular parasites. Some bacteria are also obligate intracellular parasites.

Viruses have another unifying feature in addition to their life cycles (which vary considerably but all depend on parasitism). Mature viruses have only one kind of genetic material: DNA or RNA, but not both. DNA viruses generate RNA transcripts and RNA viruses generate DNA intermediaries, but mature virions (individual virus particle) have only one or the other. There are single-stranded RNA viruses, double-stranded RNA viruses, single-stranded DNA viruses, and double-stranded DNA viruses (Table 8).

The viral genome is surrounded by a protein coating or shell, called the capsid. The genome plus the capsid is called the nucleocapsid. Some viruses consist only of naked nucleocapsids while others are surrounded by an "envelope" composed of lipids (fats) and

Table 8. Single- and Double-Stranded RNA and DNA Viruses

Examples of Single-Stranded RNA Viruses	Human immunodeficiency virus (HIV); hepatitis C virus; rhinovirus (causes common cold); coronavirus (causes SARS)
Examples of Double-Stranded RNA Viruses	Reovirus; Colorado tick fever virus
Example of Single-Stranded DNA Virus	Parvovirus
Examples of Double-Stranded DNA Viruses	Herpes simplex virus 1 and 2; poxvirus

glycoproteins (molecules composed of sugars plus proteins). An intact viral structure that has the ability to infect is commonly referred to as a virion.

Different viruses complete their life cycles within their host cells in different ways. The term "host" when used biologically generally refers to a parasitic relationship. Parasitic viral infection is not desirable to the host, and so the host cell would rather be rude and kick out the virus. Some viruses infect a cell and go about the business of making progeny virus. They do this by directing the cell's transcription and translation "machinery" to recognize the genetic material of the virus (RNA or DNA) and do its bidding. The "bidding" of that genetic material is the synthesis of the viral proteins encoded in the viral RNA or DNA. The infection proceeds such that more viral proteins and nucleic acids are made. Ultimately, these viral subunits made by the cell, under the direction of the virus, accumulate to form progeny virus.

The release mechanism from the cell can be violent and catastrophic for the cell or it can be more protracted. Some viruses accumulate in such large numbers that the cell ultimately swells and bursts (the technical term for burst is "lyse"), releasing thousands to millions of new, infectious viruses that then go on to repeat this cycle in fresh meat (uninfected cells). Some viruses, principally those that are naturally enveloped (a fatty, more or less circular covering around the outside of the virus) accumulate progeny virus within the infected cell more slowly and actually "bud" in relatively smaller numbers from the outer cell membrane. Some viruses like their new homes (infect); unpack (shed their fatty, protein coats); move in (insert their genetic material into the genome of the host cell); freeload (direct the cell to perform functions necessary to maintain the viral genome in the host genome without destroying the host cell); and ultimately are not embarrassed about deserting a sinking ship like any rat (a variety of things can cause an embedded virus, called a provirus, to "decide" it's time to leave; the virus then leaves the so-called latent phase and enters a more active phase where new viral particles are made and leave the cell, which may or may not be killed in the process).

Because viruses are so small and simple, decades of research into how they work have provided a rich source of information about not only virology but also basic genetics and the biochemistry of DNA and RNA replication and protein synthesis [☞ "Bacteriophage"].

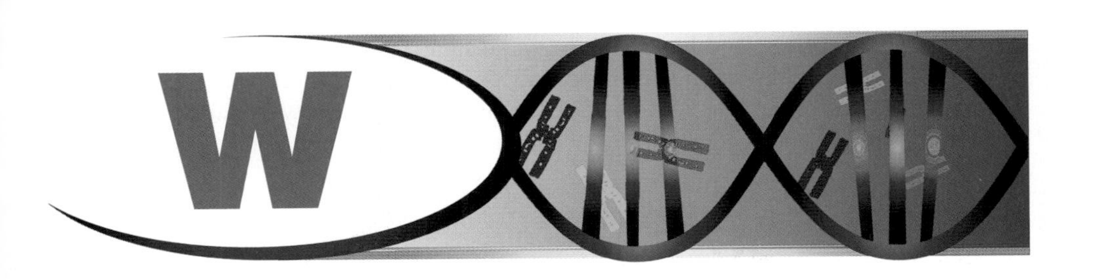

Watson, James (1928–)

Drs. James Watson and Francis Crick (with more than a little help from their professional colleagues, principally Drs. Rosalind Franklin and Maurice Wilkins) deduced the double helical nature of DNA and published it in the scientific journal, *Nature*, in 1953, as the article "Molecular Structure of Nucleic Acids."

Watson and Crick realized how that structure could be used to replicate the molecule and shared the 1962 Nobel Prize for their work (with Wilkins; Franklin had died in 1958 and the Prize is not awarded posthumously). Each went on to continued success in science, including publication, faculty positions, and acting as leaders of important scientific research institutes. Watson led the Human Genome Project at one point. He retired (amidst controversial remarks) from his last post as Director, President, and Chancellor at Cold Spring Harbor Laboratory, Long Island, NY, in late 2007 and is currently an advisor for the Allen Institute for Brain Science in Seattle, WA. For their work, scientists affectionately call one strand of double-stranded DNA the "Watson strand" and the other the "Crick strand."

A quote from Watson's and Crick's 1953 *Nature* paper (followed two pages later by Franklin's article, "Molecular Configuration in Sodium Thymonucleate [DNA]") is famous for its cheekiness and understatement, and for eventually proving to be a correct prediction: "It has not escaped our notice that the specific pairing we have postulated immediately suggests a possible copying mechanism for the genetic material."

Wild type

When molecular biologists or geneticists discuss wild-type genes, they are referring to the gene sequence that codes for the protein that has "normal" function.

An example of a wild-type protein is normal hemoglobin, while an example of the non-wild-type (mutant) version would be sickle cell hemoglobin or hemoglobin S. Individuals may carry either one or two wild-type alleles (gene forms present on the two chromosomes); when they carry two wild-type alleles, they are called homozygous wild type; when they carry one wild-type allele and one mutated version, they are called heterozygous; two mutant alleles result in the homozygous mutant genotype [☞ "Heterozygote;" "Homozygote;" "Allele"].

Wilkins, Maurice (1916–2004)

Dr. Maurice Wilkins is the often forgotten third recipient of the 1962 Nobel Prize in Physiology and Medicine for the discovery of the structure of DNA. His work, along with that of his colleague, Dr. Rosalind Franklin, was a significant contributing factor that ultimately led Drs. James Watson and Francis Crick to the realization that DNA formed a double-helical structure. Dr. Wilkins was a member of the faculty at King's College in London, where he spent almost his entire career, until his death in 2004.

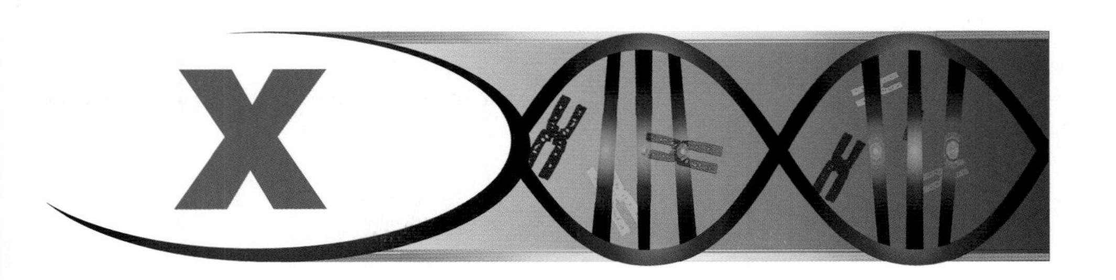

X chromosome

We, the authors, each have one son and one daughter (we think this is pretty convenient for the description in this entry). Our sons, Joshua and Ian, each have one X chromosome in their cells, just as all normal males do. Our daughters, Haley and Renee, each have two X chromosomes in their cells, like all normal females. (When they were much younger, we became very familiar with children's books that attempt to teach the alphabet. Haven't you ever noticed that they always struggle for an "X" entry? It's usually "xylophone" or "X-ray" and beyond those two it's always a reach. We got lucky here because X chromosome is an important entry for this book.)

The X chromosome is one of two chromosomes called the sex chromosomes (the non-sex chromosomes are called autosomes; the other sex chromosome is the Y chromosome and is only found in males). In normal females, two X chromosomes are present in each cell. It was initially believed that having two X chromosomes resulted in a "double-whammy" of X-linked gene expression and that this double gene expression was responsible for the "female" phenotype [☞ "Phenotype"]. It is now known, however, that this is not true. Females do not express twice as much of the proteins encoded by genes on the X chromosome as males, who have only one X chromosome per cell. In 1961, Dr. Mary Lyon hypothesized that one X chromosome per female cell is inactivated or shut down, precisely to avoid this problem. This hypothesis turned out to be correct and is called X chromosome inactivation. It is also known as Lyonization [☞ "Y chromosome"].

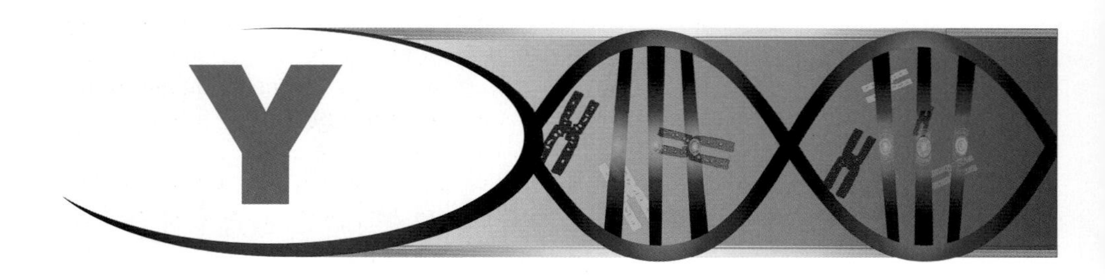

Y chromosome

The Y chromosome is the (much) smaller of the two sex chromosomes (the X chromosome is the other and larger one). It is the Y chromosome that makes the male a male.

Male gender (sex) is determined by the appropriate expression of key genes, the interaction of genes and gene products with different proteins including hormones, and a key gene present on the Y chromosome in mammals (including humans). That gene is called the *SRY* gene (for sex-related Y). Sex differentiation in utero is a complicated, intricate weave of gene expression and the influence of hormones. Even so, a single point mutation [☞ "Mutation"] in the *SRY* gene can cause an XY individual, who would normally have a male phenotype, to have an incomplete female phenotype [☞ "Phenotype"]. Some have suggested that certain characteristics are associated with *SRY*, including but not limited to refrigerator blindness and channel-surfing, two afflictions from which one of these authors in fact suffers.

Testicular feminization is an abnormal condition in individuals who are XY and who, therefore, should have a normal male phenotype. These individuals cannot properly metabolize the male hormones in the relevant cells that depend on them for correct maturation. These individuals present outwardly (phenotypically) as female [☞ "X chromosome"].

Table 9 presents chromosomal patterns and their related phenotypes.

Table 9. Chromosomal Patterns, Labels, and Phenotypes

Chromosomal Pattern*	Label	Phenotype
XX	Female	Female
XY	Male	Male
XO (one X chromosome and no Y chromosome)	Turner Syndrome	Female
XXX	Triple X Syndrome (asymptomatic)	Female
XXY	Kleinfelter's Syndrome	Male
XYY	XYY Syndrome (asymptomatic)	Male

*Additional chromosomal patterns exist.

YAC

Yeast artificial chromosome; we wouldn't have bothered you with this one but we needed another entry under "Y." You know those large, woolly animals at the zoo that look something like a cross between a woolly mammoth and a cow? Those are yaks. YACs, on the other hand, are cloning vectors used in the DNA research laboratory.

A cloning vector is a piece of DNA that can be replicated *en masse*. By using a cloning vector, such as a YAC, scientists can generate lots of copies of a specific piece of DNA. They do this by inserting that piece of DNA into the cloning vector and applying the equivalent of a molecular photocopying machine. Plasmids are another popular piece of DNA used as cloning vectors that replicate to high copy numbers in bacteria.

YACs replicate to high copy numbers inside yeast cells. Both yeast and bacteria are life forms that we can grow and multiply in the laboratory. As they increase in number, so do the cloning vectors (with the inserted piece of DNA of interest) inside them [☞ "Genetic engineering;" "Plasmid;" "Recombinant DNA;" "BAC;" "HAC;" "Vector"].

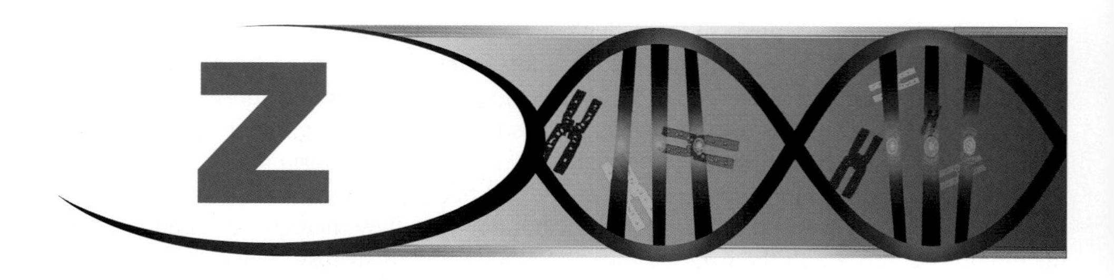

Z-DNA [☞ "A-DNA" and "B-DNA"]

Naturally occurring DNA is called B-DNA and has a right-handed turn (like an ordinary wood or metal screw) to the double helix. Another form of DNA has been observed; it has a left-handed turn to its helical structure. There is a lot of physical chemistry (with which we won't bore you) involved in the explanation for why the nucleotides within Z-DNA course through the helix in a sort of zig-zag manner. The zig-zag is why it has been termed Z-DNA.

Zygosity

Throughout this book we have made references to zygosity without clearly defining it. We have mentioned that some individuals are homo*zygous* and some are hetero*zygous*-both of these are forms of zygosity. Zygosity refers to the alleles or genes on the matching pair of chromosomes that are present in all of the body's somatic cells. When the alleles of the genes are identical, or produce the same phenotype, they are called homozygous. When they are not identical, or produce different phenotypes, they are called heterozygous [☞ "Heterozygote;" "Homozygote"].

DNA by the Numbers

Number of genes (Table 10)

Clearly an organism's complexity does not correlate with gene number (compare humans to mustard weed or rice [Table 10]). So what explains how *Homo sapiens*, clearly a complex organism, can flourish as a species with so relatively few genes?

The answer lies in the fact that we have evolved to take advantage of plasticity and modularity in our genomes. A single sequence of DNA can be spliced in multiple ways,

Table 10. Number of Genes by Organism

Organism	Estimated # of Genes	Comment
Homo sapiens	~20,500–24,500	Human beings
Mustard weed	27,170	More genes than humans
Drosophila melanogaster (fruit fly)	13,647	An excellent tool for genetic research; Groucho Marx said, "Time flies like an arrow.... fruit flies like a banana."
Saccharomyces cerevisiae	~6,000	Baker's yeast; another excellent tool for genetic research
Human immunodeficiency virus (HIV-1)	9	Like all viruses, HIV is an obligate intracellular parasite.
Human mitochondrion	37	There is evidence that this cellular organelle is an ancient pathogen that has become part of us through human evolution.
Chlamydia trachomatis	936	Most commonly sexually transmitted bacterium in the U.S.
Mycobacterium tuberculosis	3,959	Bacterial pathogen that causes tuberculosis
Mycobacterium leprae	1,604	Bacterial pathogen that causes leprosy
Agrobacterium tumefaciens	5,419	This bacterium causes (crown gall) tumors in plants via a *Ti* (tumor-inducing) plasmid that incorporates itself into the plant genome; its discovery helped establish the field of plant genetic engineering.
Yersinia pestis	4,052	Bacterial cause of bubonic plague
Oryza sativa	~60,000	Rice

such that more than one RNA transcript and ultimately more than one protein may be encoded by one DNA sequence. Genes can rearrange themselves to generate multiple transcripts and proteins from one stretch of DNA. Genes involved in generating proteins responsible for the immune response behave in this way.

Remember that it is ultimately proteins that carry on the "business of life" inside the cell and throughout the body. So while there may only be about 20,500–24,500 relatively static (in number) human genes, the proteome, or complete complement of proteins in the body, is much larger and much more dynamic in terms of which proteins are present in which cell at any given point in time. From this it also follows that the transcriptome, the full complement of transcription products encoded in our DNA, is also more complex than the genome.

4-3-20

The molecular trifecta. This is the numerical equivalent of the central dogma of molecular biology, a term coined by Francis Crick to depict the flow of genetic information from DNA to RNA to protein (since adapted for the way retroviruses transmit genetic information). Four nucleotides in DNA (A, C, G, T) are read three bases at a time in RNA during translation [☞ "Anticodon;" "Codon;" "Translation"] to generate 20 amino acids (Table 11) that are strung together in the ribosomes into proteins that carry on the stuff of life.

Table 11. Amino Acids

Amino Acid*	Three-Letter Abbreviation	One-Letter Abbreviation	Molecular Weight	Codons (U in RNA Replaces the T Found in DNA)
Alanine	Ala	A	89.09	GCU, GCC, GCA, GCG
Arginine	Arg	R	174.20	CGU, CGC, CGA, CGG, AGA, AGG
Asparagine	Asn	N	132.12	AAU, AAC
Aspartic acid	Asp	D	133.10	GAU, GAC
Cysteine	Cys	C	121.15	UGU, UGC
Glutamine	Gln	Q	146.15	CAA, CAG
Glutamic acid	Glu	E	147.13	GAA, GAG
Glycine	Gly	G	75.07	GGU, GGC, GGA, GGG
Histidine	His	H	155.16	CAU, CAC
Isoleucine	Ile	I	131.17	AUU, AUC, AUA
Leucine	Leu	L	131.17	UUA, UUG CUU, CUC, CUA, CUG
Lysine	Lys	K	146.19	AAA, AAG
Methionine	Met	M	149.21	AUG
Phenylalanine	Phe	F	165.19	UUU, UUC
Proline	Pro	P	115.13	CCU, CCC, CCA, CCG
Serine	Ser	S	105.09	UCU, UCC, UCA, UCG, AGU, AGC
Threonine	Thr	T	119.12	ACU, ACC, ACA, ACG
Tryptophan	Trp	W	204.23	UGG
Tyrosine	Tyr	Y	181.19	UAU, UAC
Valine	Val	V	117.15	GUU, GUC, GUA, GUG
STOP	TER			UAA, UAG, UGA

*The amino acids listed in "boldface" are the "essential amino acids" that cannot be synthesized by the body.

The STOP or TERMINATION (TER) codons listed in the last row are the ones that signal the protein-synthesizing machinery of the cell to end translation and terminate the growing polypeptide (protein) chain. AUG, on the other hand, codes for methionine and initiates the process; it is termed the initiation codon. So the genetic code has some punctuation akin to capital letters at the beginning of a sentence and periods at the end.

Amino acids are made up of varying amounts of carbon, oxygen, hydrogen, and nitrogen. Cys and Met also contain sulfur. If the above table motivates those high school and college students reading this to go to a freshman chemistry book and learn what "molecular weight" is, we have done our job. Note also that Leu and Ile have the same molecular weight and very similar names. The term "iso" refers to the fact that isoleucine is an isomer (or alternative form) of leucine, with all the same molecules, hence the identical molecular weight, but in a slightly different configuration.

Human beings must ingest nine so-called essential amino acids that cannot be synthesized by the body. The essential amino acids are listed above in boldface. Note that Cys can partially substitute for Met, because they both contain sulfur, and that Tyr can partially substitute for Phe.

For any given amino acid, note that many (not all) of the first two bases in a triplet codon are identical while the third base tends to "wobble" yet still allow coding for the same amino acid. For example, CCU, CCC, CCA, and CCG all code for proline and the only difference among the four codons is in the third base. Francis Crick referred to this as "wobble at the third codon position." The advantage of "wobble" is that tRNA can dissociate from mRNA more quickly without generating an error, thereby helping to accelerate protein synthesis.

Another take-home message from the fact that many amino acids are encoded by more than one codon is the so-called redundancy of the code.

The cracking of the genetic code was Nobel Prize-winning work. The prize in Physiology or Medicine in 1968 was shared equally in recognition of the work by Marshall W. Nirenberg, Robert W. Holley, and Har Gobind Khorana. Using a cell-free protein-synthesizing system, these investigators showed that polyuridylic acid (essentially a long tract of uridines equivalent to UUU in the table above) generated phenylalanine. Manipulation of the sequence of the input RNA translated by the cell-free synthesizing system resulted in different amino acid outputs, which ultimately led to cracking the entire code.

High school and college students reading this might consider that it may (or may not) do more for your future career if you memorized the above table instead of memorizing a headful of useless baseball statistics like 755 (the number of home runs Hank Aaron hit in his career) or 369 (the number of times Joe DiMaggio struck out in his career). If you infer from this remark that one of your authors is a highly frustrated baseball (NY Mets) fan who has not followed his own advice, you'd be correct.

5%

Only about five percent of the human genome is transcribed; the total collection of transcripts in the cell is called the transcriptome. Because each gene may produce differing transcripts, the transcriptome is more complex than its coding DNA.

21mer

21 bases happens to be a very good length for probes and primers (and is also a very good number at the Black Jack table). With respect to DNA, "21mer" refers to a stretch of synthetically prepared DNA 21 bases in length. There's no reason you couldn't have a "16mer" or a "37mer," or the like. It turns out that 21mers are a popular length for primers and probes [☞ "Primer;" "Probe"] because for physical and chemical reasons, these form nicely stable and specific hybrids to target sequences of DNA that are perfectly complementary to them [☞ "Complementary strands of DNA"].

23

The number of human chromosome pairs; normal individuals have 23 pairs in their somatic cells: pairs 1–22 plus an XX (female) or XY pair. Table 12 lists individual chromosome statistics.

Note that the Y chromosome, found only in males, has the fewest genes (~230). Please feel free to make up your own joke. There are 70 chromosomes in the camel genome, 78 in the chicken, and 34 in the porcupine. How do we know? Check out this URL: http://morgan.rutgers.edu/morganwebframes/level1/page2/ChromNum.html.

Table 12. Chromosomes, Genes, and Associated Diseases or Characteristics

Chromosome #	# of Genes*	# of Mb**	Some Associated Diseases or Characteristics***
1	2968	279	Colon and breast cancer; deafness
2	2288	251	Programmed cell death (☞ apoptosis); cataracts; cleft palate; red hair
3	2032	221	Night blindness; short stature
4	1297	197	Another gene for red hair color; acute myeloid leukemia; phenylketonuria
5	1643	198	Obesity with impaired prohormone processing; diphtheria toxin receptor; susceptibility to attention deficit hyperactivity disorder
6	1963	176	Dyslexia; Celiac disease (gluten insensitivity); susceptibility to coronary artery disease
7	1443	163	Hereditary pancreatitis; susceptibility to ulcerative colitis; growth hormone deficient dwarfism
8	1127	148	Opiate receptor; human papilloma virus type 18 (causes cervical cancer) integration site; congenital adrenal hyperplasia
9	1299	140	X-ray damage repair; fructose intolerance; melanoma; car diomyopathy
10	1440	143	Polycystic kidney disease; Warfarin sensitivity; glaucoma
11	2093	148	Insulin-dependent diabetes mellitus; sickle cell anemia; osteoporosis
12	1652	142	Acute alcohol intolerance; taste receptors, familial Alzheimer's disease
13	748	118	Non-small-cell lung cancer; bladder cancer; schizophrenia susceptibility

Table 12. Chromosomes, Genes, and Associated Diseases or Characteristics *(Continued)*

Chromosome #	# of Genes*	# of Mb**	Some Associated Diseases or Characteristics***
14	1098	107	Atypical Marfan syndrome; Graves disease; congenital hyperthyroidism
15	1122	100	Severe mental retardation; Tay-Sachs disease; brown eye and hair color; colorectal cancer
16	1098	104	Familial gastric cancer; cocaine and antidepressant sensitivity; familial mitral valve prolapse
17	1576	88	Neuroblastoma; HER-2/*neu* (breast cancer Tamoxifen candidates); delayed progression of HIV disease
18	766	86	Familial carpal tunnel syndrome; Paget disease of bone; hepatitis B virus integration site
19	1454	72	Green/blue eye color; late onset Alzheimer's disease; prostate-specific antigen (a prostate cancer marker)
20	927	66	Gigantism; hemolytic anemia; susceptibility to myocardial infarction (heart attack)
21	303	45	Amyotrophic lateral sclerosis (Lou Gehrig's disease); influenza resistance
22	288	48	Chronic myeloid leukemia; cat eye syndrome
X	1184	163	Becker and Duchenne muscular dystrophy; fragile X mental retardation; color blindness
Y	231	51	Sex-determining region (Y); azoospermia factors

*approximate

**Mb = megabases or one million bases

*** The listing is not meant to suggest that the diseases and characteristics listed are necessarily caused by mutations in but a single site on a single chromosome; most diseases are caused by many mutations and different and complex factors; the listing is meant to be illustrative, not exhaustive, with respect to pathogenesis or cause of the characteristic. For example, using the website below, one will find gene defects associated with, for example, breast cancer, colorectal cancer, leukemia, and deafness, on many of the chromosomes.

SOURCE: http://genome.gsc.riken.go.jp/hgmis/posters/chromosome/faqs.html

26

As of spring 2008, there were 26 U.S. institutions offering training to MDs in molecular pathology; learn more at the website of the Association for Molecular Pathology (www.amp.org).

31

As of spring 2008, there were 31 Masters-level genetic counseling programs in North America with either Full, Interim, or Provisional accreditation (or Recognized New Program Status) that have met or are in the process of satisfying the rigorous accreditation criteria established by the American Board of Genetic Counseling (www.abgc.net).

46

The number of chromosomes in a human cell. *Homo sapiens* (humans) have 22 pairs of autosomes (non-sex chromosomes) in each cell (except red blood cells) and one pair of sex chromosomes for a total of 23 pairs (46 in all). Females have one pair of sex chromosomes (two X chromosomes); males also have one pair of sex chromosomes (one X and one Y chromosome).

50

It took 50 years, a very short time in the history of humankind, to progress from the discovery of the DNA double helical structure to the completion of the sequencing of its three billion base pairs. As the rate of progress in medical science and genomics continues to quicken due to new discoveries and new technologies, imagine what progress genomics will have brought to bear on the medical sciences that will be practiced on those born in 2003 and alive in 2053.

97

The percentage of insurance claims currently being reimbursed in the U.S. for genetic breast cancer predisposition testing (source: Bishop JM. How to Win the Nobel Prize. Cambridge, MA: Harvard University Press, 2003).

98

The percent homology between the human and chimpanzee genomes; apparently two percent makes a big difference.

99.9

The percent homology between any two human's genomes; even identical twins exhibit differences at the DNA sequence level [☞ "Copy number variation"].

379

This is the length, in bases, of mitochondrial DNA (mtDNA) that a scientific team was able to string together from the arm bone of a recovered skeleton of a Neanderthal. By examining this 379 base pair DNA fragment, the German team was able to show that modern day *Homo sapiens* did not descend from Neanderthals, but rather, Neanderthals were a separate species that became extinct. They were able to demonstrate this due to the significant differences in the mtDNA between the two species. This finding shows the remarkable general stability of DNA over time and the power of PCR to generate copies of these DNA fragments for study.

1,000–1,000,000

The smallest genes are about one thousand nucleotide bases long; the longest about a million.

1,546

As of March 2008, 1,546 diseases were listed at www.genetests.org/ and tested for at more than 600 different clinical and research laboratories.

1953

This is the year James D. Watson and Francis H. C. Crick published their elucidation of the structure of DNA. The article was called "Molecular Structure of Nucleic Acids: A Structure for Deoxyribose Nucleic Acid" and was published in the British journal, *Nature*, volume 171, page 737, April 25, 1953. The paper was followed quickly by another Watson and Crick manuscript in which they more fully described the replication process for DNA: "Genetical Implications of the Structure of Deoxyribonucleic Acid," published in *Nature*, volume 171, pages 964–967, May 30, 1953.

1958

The year one of your author's genetic material was expressed (actually, since DHF was born in May 1958, gene expression began in utero in 1957).

1995

The year that Craig Venter's team at The Institute for Genome Research completed the first sequencing of a free-living (i.e., not a parasite, like a virus,) organism, the bacteria *Haemophilus influenza. H. flu*, as it's called, is a human pathogen that causes, among other things, respiratory infections and, left untreated, can prove fatal. Its genome is 1.83 Megabases (Mb) in length or 1,830,000 bases. The human genome is about 1600 times larger, and enhancements to technology allowed the completion of the sequencing of the human genome eight years later.

2003

The year in which the sequencing of the human genome was completed. There is symmetry in this accomplishment as 2003 was the 50th anniversary of Watson's and Crick's elucidation of the double helical structure of DNA.

16,569

The number of base pairs in the circular DNA molecule contained in our cells' mitochondria (distinct from the much larger DNA genome in our nuclei).

20,500, "give or take"

The best estimates, at the time this book was written, for the number of genes humans have is 20,500. This number represents the number of open reading frames [☞ "ORF;" under the heading of "if it looks like a duck and quacks like a duck, it must be a duck," so, too, it is with ORFs and genes], and so this is the number of genes ascribed to the human genome. This does not mean, however, that every human has 20,500 or 20,311 or 20,984 genes (whatever the final number may turn out to be). Indeed to suggest a number is somewhat irrelevant because of something known as copy number variation. Genes can be deleted or duplicated (or amplified to a number even greater than two) among individuals. Thus, there is variation in the actual number of genes between any two individuals; estimates range from 60 to 100 gene number differences [☞ "Copy number variation"].

10,000,000

There are about 10,000,000 single nucleotide polymorphisms (SNPs) that differentiate humans from each other within the larger three billion base pair genome. SNPs are useful as markers of disease, disease susceptibility, and response to therapeutic drugs, among other medically important characteristics.

3,000,000,000 (3×10^9)

The number of nucleotide base pairs in a human sperm or egg cell. Non-sex cells (somatic cells) like stomach, nerve, and muscle cells have twice as much DNA. Here are some fairly useless arithmetic facts but they serve to illustrate just how much DNA is in our cells and how tightly packaged it is.

If one multiplies the number of base pairs in a somatic cell (3×10^9) by the length of the DNA purified from a <u>single</u> cell, and if you could string out that DNA in a straight line (3.4×10^{-10} meters per base pair), the product is about 1.02 meters of DNA per cell. That's more than three feet of DNA in one cell.

If you multiply 1.02 meters of DNA in one cell by the number of cells in a mature adult human (about 3.5×10^{13}), the product is 3.57×10^{13} meters.

It's about 93,000,000 miles from the Earth to the Sun (one way). One mile is about 1600 meters so the distance to the Sun is about 1.49×10^{11} meters. If you divide 3.57×10^{13} (the number of meters of DNA in one person) by 1.49×10^{11}, you learn that the amount of DNA in just one human being could be strung back and forth between the Sun and the Earth about 240 times. That's a lot of DNA!

Professional Organizations

Following is a list of professional societies and organizations (1) whose membership is involved in medical research and in the implementation, certification, education, training, and proper use of DNA technology in the clinical laboratory, or (2) that are otherwise involved with DNA; it is *not* a comprehensive list.

AAAS	American Association for the Advancement of Science, Washington, DC
AABB	American Association of Blood Banks, Bethesda, MD
AACC	American Association for Clinical Chemistry, Washington, DC
ACMG	American College of Medical Genetics, Bethesda, MD
AMP	Association for Molecular Pathology, Bethesda, MD
APSMV	Asia Pacific Society for Medical Virology, offices listed by nation
ASCP	American Society of Clinical Pathologists, Chicago, IL
ASHG	American Society of Human Genetics, Bethesda, MD
ASM	American Society for Microbiology, Washington, DC
ATCC	American Type Culture Collection, Rockville, MD
CAP	College of American Pathologists, Northfield, IL
DOE	Department of Energy, Washington, DC
EMBO	European Molecular Biology Organization, Heidelberg, Germany
ESCMID	European Society of Clinical Microbiology and Infectious Diseases, Basel, Switzerland
ESCV	European Society for Clinical Virology, offices listed by nation
FBI	Federal Bureau of Investigation, Washington, DC, and Quantico, VA
FEMS	Federation of European Microbiological Societies, offices listed by nation
HUGO	Human Genome Organization (international)
IUMS	International Union of Microbiological Societies, Utrecht, The Netherlands
IUPAC	International Union of Pure and Applied Chemistry, offices listed by nation
MRC	Medical Research Council (UK)
NCA	National Certification Agency for Medical Laboratory Personnel, Lenexa, KS
NCI	National Cancer Institute, Frederick, MD

NHGRI	National Human Genome Research Institute, Bethesda, MD
NIAID	National Institute of Allergies and Infectious Disease, Bethesda, MD
NIDDK	National Institute of Diabetes and Digestive and Kidney Diseases, Bethesda, MD
NIH	National Institutes of Health, Bethesda, MD
PASCV	Pan American Society for Clinical Virology, offices listed by nation
USCAP	United States and Canadian Academy of Pathology, Augusta, GA

The Human Genome Project

Information about the human genome project and associated research may be viewed at www.genome.gov.

The National Cancer Institute

The National Cancer Institute (NCI) homepage (www.cancer.gov) is useful in learning about cancer prevention, diagnosis, treatment, and cancer therapy clinical trials conducted by NCI-sponsored researchers, pharmaceutical companies, and international groups.

Further Reading

If this book has whetted your appetite for more serious and intensive treatment of these subjects, there is a wealth of further information available. For Internet browsers, there's lots to review. Just type in key words like "DNA," "Genetics," or "Molecular Pathology" in your search and you'll be amazed at how much information pops up.

For those of you who'd like to do further reading, there is a list of books on the Association for Molecular Pathology's website (www.amp.org). Go to the site, click on "About AMP," then "Committees," then "Training and Education," then "Educational Materials," and then "Molecular Pathology Booklist."

Special mention must be made of *The Eighth Day of Creation: Makers of the Revolution in Biology* by Horace Freeland Judson, last published in 1996. This is a monumental and rich description of the history of molecular biology. It's not beach reading, but is highly recommended to anyone interested in the subject. The ISBN # is 0879694785 and the book is available online for about $40.

Here are six more recommended books of general interest that are available for purchase on the Web:

- *What Mad Pursuit: A Personal View of Scientific Discovery*, by Francis Crick. (Published in 1990; ISBN #: 0465091385)
- *The Double Helix: A Personal Account of the Discovery of the Structure of DNA*, by James Watson. (Last published in 2001; ISBN #: 074321630X)
- *Rosalind Franklin: The Dark Lady of DNA*, by Brenda Maddox. (Published in 2002; ISBN #: 0060184078)
- *DNA: The Secret of Life*, by James D. Watson and Andrew Berry. (Published in 2003; ISBN # 0375415467)
- *How to Win the Nobel Prize*, by J. Michael Bishop. (Published in 2003; ISBN # 0674008804)
- *Genome: The Autobiography of a Species in 23 Chapters*, by Matt Ridley. (Published in 2006; ISBN # 9780060894085)

About the Authors

Daniel H. Farkas, PhD, HCLD, CC, CLSp(MB), FACB

Daniel H. Farkas is Executive Director of the Center for Molecular Medicine in Grand Rapids, MI, where he has established an advanced molecular pathology laboratory in support of 21st century diagnostic applications and clinical trials for both the diagnostics and pharmaceutical industries. The Center is a joint venture between Grand Rapids-based Spectrum Health and the Van Andel Institute and represents a new paradigm in molecular diagnostics.

Dr. Farkas was Director of Molecular Pathology at The Methodist Hospital (TMH) in Houston from 2002 to 2005, where he established a new hospital-based molecular diagnostics service for the third time in his career. While at TMH, Dr. Farkas was Associate Professor of Pathology and Laboratory Medicine at Weill Medical College of Cornell University, New York City (2005), and Associate Professor of Pathology at Baylor College of Medicine, Houston (2002–2004). Dr. Farkas has done two stints in the biotech industry, most notably from 1998 to 2002, when he was Director of Clinical Diagnostics at Clinical Micro Sensors (which became a Motorola company in spring 2000) and played a significant role in the ultimate FDA approval of a DNA-chip-based test for cystic fibrosis mutation detection. Dr. Farkas established and was co-director of the Molecular Probe Laboratory at William Beaumont Hospital in Royal Oak, MI, from 1991 to1998. Prior to that, in 1989, he established the Diagnostic Molecular Pathology Laboratory at Saint Barnabas Medical Center in Livingston, NJ. Dr. Farkas is currently on the faculty at his alma mater, Michigan State University. He also served on the faculty of the Beaumont Hospital Medical Technologist Training Program, School of Medical Technology.

Dr. Farkas has published widely in molecular diagnostics and, as a recognized expert, has lectured internationally on the subject for 20 years. Active in many industry associations, he served on the College of American Pathologists Molecular Pathology Resource Committee for nine years, and has served on the Board of Directors (2007–2009) of the American Association for Clinical Chemistry and as President (2003) of the Association for Molecular Pathology (AMP), from which he later received, in 2007, the AMP Leadership Award. Dr. Farkas serves on the editorial boards of the *Journal of Molecular Diagnostics*, *Biotechnology Healthcare*, and *Diagnostic Molecular Pathology*. He is a fellow of the National Academy of Clinical Biochemistry, and was a member of (and is now consultant to) the FDA Clinical Molecular Genetics Advisory panel. As the first person to be certified by the American Board of Bioanalysis (ABB; Board member 2004–2009) in Molecular Diagnostics, Dr. Farkas holds the credentials of High Complexity Clinical Laboratory Director (HCLD) and Clinical Consultant (CC) from ABB, as well as Certified Laboratory Specialist in Molecular Biology from the National Credentialing Agency for Laboratory Personnel.

Dr. Farkas, an avid reader and photographer, lives with his wife, Becky; his son, Josh; and his daughter, Haley, in Washington, MI (outside Detroit) with their Cavalier King Charles Spaniels, Peanut and Cosmo. He still roots (sometimes in private misery, sometimes in unmitigated joy) for the New York Mets, Jets, Rangers, and Knicks—and the Michigan State Spartans.

Carol A. Holland, PhD, MT (ASCP)

Carol A. Holland currently works at Beckman Coulter, Inc., in Technical Marketing. Dr. Holland was born in Detroit, MI, and grew up (quite often literally) in Michigan's Wolverine Lake. She earned both a Bachelor's and Master's degree in Biological Sciences from Oakland University, Rochester, MI. During this time, Dr. Holland worked her way through college as a laboratory aid, Medical Laboratory Technician (ASCP), and a Medical Technologist (ASCP) at St. Joseph Mercy Hospital in Pontiac, MI. She then attended Wayne State University in Detroit where she earned her doctorate in Molecular Biology in 1996, followed by a Postdoctoral Fellowship in Medical Microbiology at Henry Ford Health Systems, also in Detroit.

After Dr. Holland completed her fellowship, she joined the Infectious Disease Department at Henry Ford Health Systems, where she performed research on the hepatitis C virus and HIV, until she was recruited to be the Technical Director of Molecular Pathology at William Beaumont Hospital in Royal Oak, MI. In 2006, to broaden her diagnostics experience, she moved to Beckman Coulter Inc.'s High Sensitivity Testing Group.

Dr. Holland is adjunct associate professor at Oakland University, has served as a consultant for several diagnostic companies, and serves on the Editorial Board for the *Annals of Clinical & Laboratory Science*. An active member of the Association for Molecular Pathology, the American Association for Clinical Chemistry, the American Society for Microbiology, and the Association of Clinical Scientists, Dr. Holland has published numerous articles and book chapters and has a patent pending for a molecular diagnostic test for the hepatitis C virus (HCV).

Dr. Holland loves all water sports, especially swimming. She lives with her husband, Phil Cunningham, a professor at Wayne State University, and son Ian in Troy, MI. She also has a daughter, Renee; son-in-law, Jason; and a delightful grandson, Griffin, who live in the mountains of Colorado.

Index

Page numbers in italics indicate that the information can be found in figures or tables. Page numbers in bold type identify an encyclopedia entry.